the ATLAS of a CHANGING CLIMATE

OUR EVOLVING PLANET
VISUALIZED WITH MORE THAN 100
MAPS, CHARTS, AND INFOGRAPHICS

the ATLAS of a CHANGING CLIMATE

BRIAN BUMA

TIMBER PRESS
PORTLAND, OREGON

This book is dedicated to the artist that scratched a crude mountain onto a mammoth tusk in the Czech mountains; to Alexander von Humboldt, who brought the unity of nature to the world; and to the modern data wizards that distill our complex world into visuals that communicate global information, trends, and urgency in equal measure. Artistry in science matters.

It is also dedicated to the environmental activists that tirelessly fight to contain and mitigate the damage we casually inflict on the natural world. The charts and maps in this volume in some ways document a decline, but those still fighting have not given up hope.

Finally, it is dedicated to my family—Audrey, Cole, and Owen (and Blue the dog, Carbon the black cat, and our chickens), who put up with me monopolizing our only spare bedroom to type, retype, and type again while COVID shut down the rest of the world.

Photography and illustration credits appear on page 259.

Published in 2021 by Timber Press, Inc.
The Haseltine Building
133 S.W. Second Avenue, Suite 450
Portland, Oregon 97204-3527
timberpress.com

Printed in China
Text and cover design by Adrianna Sutton
Text is set in Cala, a typeface designed by Dieter Hofrichter in 2011
Cover illustration by Clara Prieto

ISBN 978-1-60469-994-4

Catalog records for this book are available from the
Library of Congress and the British Library.

CONTENTS

CHARTING THE NATURAL WORLD

In 2019, I stood on the peak of *Isla Hornos*, the fabled Cape Horn of sailing lore. Below me, about 1000 feet (300 m) down, ocean swells that had circled the globe without interruption exploded against the black, rocky headwall. Black and white penguins chattered, absurdly bright green bunchgrasses swayed at head height, and a lone Andean condor floated just overhead on an updraft, seemingly close enough to touch—plenty of life, despite the cold. I was there leading a National Geographic expedition to explore the island and find the world's southernmost tree (which we did—a stubby southern beech tree about waist high and quite healthy).

It was the winds, though, that monopolized my thoughts. Long walls of clouds screamed past overhead, and not scattered white things, but whole weather systems. They weren't swirling, or gently billowing, or any of the normal things clouds do on a summer day, but rather rocketing above, aloof and hard. I had a strong, disorienting perspective: the island was a pebble at the bottom of a creek with water coursing above and taking no notice of the small bumps and dimples on the creek bed.

The indifferent currents in the sky frequently reached hurricane force, and when they touched down, they immediately set about flattening tents and knocking unsuspecting scientists into the heath. There's nothing like being slammed to the ground to bring one's perspective back to earth! We were insects under the waves, insignificant to an atmospheric system global in size and scope. The wind did not notice the island, nor the people struggling on it. It was too big, we were too small. As I lay back and watched each concussive swell of storm flow above, two contrasting viewpoints jockeyed back and forth. There was the aforementioned insignificance, being dwarfed by a literal planet-sized system, revealed above me as nowhere else on Earth. But I also felt connected by that exposure, part of the system itself. A small part, to be sure, but with nothing to disconnect me from the atmosphere, the feeling was inescapable. Engulfed by the wind, truly a part of the world. It was impossible to forget that while any single breath of mine was being ripped away by the gale on an eastward journey around the world, joining the albatrosses, that same breath would come back, eventually, from the other direction.

———

You and I both have a fundamental challenge with understanding the natural world: scale. By that I mean two things. First, the processes that govern Earth's systems operate at the spatial scale of the whole planet—far different from our human experience, which is limited to our neighborhoods and immediate surroundings. Second, the temporal or time-related scale on which these global systems operate is also very divorced from our perspective; our calendars, broken into weeks and months, poorly prepare us to conceptualize variability and predictable patterns over centuries or millennia—many lifetimes to us, but a single cycle to the world.

It can be a real challenge to understand the basics of the broad systems of the natural world and to get a sense of how they are changing at their proper scale. And that understanding is required to truly get a handle on change. Furthermore, variability (the temporal component) is incredibly important—a subject often lost in news about average temperatures rising or average rainfall declining. It turns out that

PREVIOUS
Southern South America is a land of high ice, windy plains, dark forests, precarious islands, and also humanity, all surrounded by the southern ocean. Cape Horn is the southernmost headland of the Tierra del Fuego archipelago of sourthern Chile, on small *Isla Hornos*, just south of the mainland's southern tip at the bottom of this image.

average changes are not always as significant as variability and changes in variability. In other words, it is not necessarily the slightly warmer temperature over decades that kills a forest, it can be only a few years of extreme heat, hotter than they used to be. Change can be sporadic and sudden, a series of thresholds that are crossed, not only a gradual process of warming average annual temperatures. It's a big world, with lots of variability in space and time. Science can quantify all this information about systems and scales (at least as far as we understand it now!), but the examples often fall flat because the scale of the issues are global and we are limited in our perspective to the local.

Moving one's perspective from the local to the global requires imagination and art as much as science. Therefore you need not just words and numbers, but imagery as well. Charts and maps illustrate what words and our personal experience often cannot—the spatial scale of the natural world, continents at a glance. Imagery spans a wide historical period: from when our understanding of how big and interlocked things really are was just in its infancy, to the modern satellite era, in which global views are as common as the evening weather report. Like snapshots, charts and maps display what we, or the artists, think is important to communicate. Sometimes they reveal the way people thought about the world; a resource to be exploited or a service to be protected, for example.

Other times, charts can serve as time capsules, bridging temporal scales, preserving a vision of what *was then* that we can compare to what *is now*. Maps and charts give perspective. The numbers of science must be paired with history and visualization to appreciate the world as it is, and to suggest where it's going.

YOU AND I BOTH HAVE A FUNDAMENTAL CHALLENGE WITH UNDERSTANDING THE NATURAL WORLD: SCALE.

———

We can put numbers to our perception of the environment: the atmosphere, waters, lands, and cities that bound our lives. We humans average a little less than 6 ft. (1.8 m) tall and can see about 3 mi. (5 km) on clear days from a level spot, which corresponds to seeing an area of about 30 sq. mi. (78.5 sq km) if you slowly pivot and look in all directions. The world is 24,901 mi. (40,075 km) around the equator, and about 197,000,000 sq. mi. (510,000,000 sq km) in area. That is a difference in scale of around 8000 times in terms of circumference and 6.5 million times (!) in terms of area. It is hard to overstate the difficulties this scale mismatch causes in understanding the world and global climate change; it is a scale wholly disconnected from our individual lives.

Prior to broadscale mapping and charts, our ability to comprehend the global system was severely constrained. Consider climate. Seasons were clear enough, certainly, as were differences between the northern and southern hemispheres. But anticipating a winter storm? Explaining a drought? We went to mysticism, curses, and disgruntled gods. Those processes, however, are not random events that occasionally sprout up out of nothing; they do not even have a specific origin. Weather emerges from a system that is continual; everything is connected to everything else in space and time. But we cannot see that from our limited viewpoint on the ground. We are constantly affected by processes that are in motion over the horizon, and we in turn trigger changes felt by those downwind from us. And it all circles around to be felt by us again. There is no independence on a spherical world. "When we try to pick out anything by itself, we find it hitched to everything else in the Universe," wrote the naturalist and explorer John Muir. He was not wrong. To understand the natural world requires a global perspective.

This is where the role of graphics comes in, why data and numbers must be accompanied by art and visualization. There is a reason "The Blue Marble," an ordinary photo snapped on December 7, 1972, is one of the most reproduced images in history. It gave us our home as a singular object, taking us from the realm of local experience to a visceral connection with the scale of the world, via photography and living color. "The Blue Marble" is Muir's quote made tangible. Maps, charts, and imagery have a way of transforming and scaling our perception like nothing else can.

An excellent example of the power of this combination of mapping-plus-scientific-investigation-plus-time is the work of Alexander von Humboldt. Humboldt, a naturalist and explorer of the late 18th and early 19th centuries, was perhaps the father of this sort of scientific mapping. A true polymath, he excelled in observation and connections. For several decades, Humboldt was one of the most famous people on the planet. Incredibly energetic (calling coffee "concentrated sunbeams") and relentlessly curious, he wrote on climate, plants, animals, adventure travel, political philosophy, and geology in equal turns; explored Latin America; hypothesized about plate tectonics 100 years before the theory was confirmed; and brought the study of the natural world to a literal "worldly" scale (an abbreviated list!). One of his more substantial achievements was the scope of his science, which displayed a recognition that everything is connected to each other—that the study of nature cannot be done piecemeal, isolated from place to place. He did so through "Naturgemälde" (roughly translated as "painting of nature" or unity of nature); illustrating the idea that to truly understand a thing, one needs to understand its context

There is no "somewhere else." The famous "Blue Marble" photo, from Apollo 17, makes that truth stark and painfully clear. Earth's system is a self-contained one, a bubble of water and air. Visualization made this scientific reality tangible to billions of people: everything is a piece of a single, and small, whole. A cyclone moves onto the Indian subcontinent; Africa peeks from below white clouds; Antarctica wreathes the bottom (actually, the photo was originally taken with Antarctica on the top; the photo was reoriented to seem more familiar).

in the world. To study individuals of a species, you needed to understand their communities. Humans were no longer a separate piece, but a component of the larger whole. This grand view, that all life is connected, made the subsequent global theories of evolution (Darwin idolized Humboldt), plate tectonics, the field of biogeography, and many other avenues of investigation feasible. This shift of perspective—coupling deep scientific study of a topic with a simultaneous look at the broad context—was revolutionary.

One of Humboldt's many famous contributions, and perhaps the best known, is his work on the massive Ecuadorian mountain known as Chimborazo, 20,549 ft. (6263 m), and other mountains in the Andes. At the time, it was generally believed that "Chimbo" was the highest mountain in the world. (The belief is not necessarily wrong. The beautifully conic stratovolcano, glacial capped, nearly straddles the equator at just one degree south latitude. As it turns out, our globe is not actually a perfect sphere but rather slightly deformed (an oblate spheroid), meaning the poles are a bit flattened and the equator bulges out due to the spinning rotation of Earth. Because of Chimbo's location right on the equator, it actually *is* the highest point if you measure from the center of the planet.) Humboldt climbed several peaks around that time looking at vegetation, and his attempt on Chimbo was the crowning achievement, even though he did not summit. The party reached about 1000 ft. (300 m) from the top before being stopped by a wall of ice; it was the highest anyone had climbed to that date. This was an exploration achievement of the highest order, but it was how he communicated his science that was revolutionary.

The *Tableau Physique*, published in 1807, communicates Humboldt's science via art and map, rather than table and graph. It revolutionized science and our perception of the natural world through its innovative data communication and visual appeal. The mountain is clearly there but scaled to a page; the vegetation zones you would experience as you climb are delineated. Whereas an ascent would take weeks, making appreciation of the patterns difficult, here the zones are all laid out so that the mountain *system*, as opposed to parts of the mountain, are visible as a unified whole. Abstraction occurs when necessary: for example, the mountain sketched is impossibly steep, the better to convey his point—how life changes as one ascends. Humboldt sacrificed topographical reality for communicative viability, and it works. Agricultural plants are represented alongside physical factors, emphasizing people and the limits of the human-social-environmental setting. The broad trends revealed by Humboldt communicate the scale of the processes of the natural world that are simply too big to appreciate from the average human experience. It is a new way of looking at the world, and a modern one—necessary to appreciate the challenges of our time.

WAIT, YOU MIGHT SAY, HISTORICAL RECORDS ARE NOT ALWAYS RELIABLE, ARE THEY?

This work is not just a historical novelty. An additional value of this work is how it captures the scale of time, which we cannot recreate any other way. The records at Chimborazo and elsewhere in Ecuador are again an excellent illustration. Back then, agriculture topped out at about 11,800 ft. (3600 m) elevation. Now potatoes grow happily at 13,000 ft. (4000 m) in Ecuador. Previously, seed plants were limited to about 15,000 ft. (4600 m), but a recent study by researchers at Aarhus University found them at 17,000 ft. (5185 m). This is all evidence that as the climate warms, the environment responds, and in a big way. The temporal scale of climate change, over centuries, can be just as challenging to visualize as big spatial processes like climate itself—but with the right perspective, it becomes possible.

Wait, you might say, historical records are not always reliable, are they? That can be true, and in this case, it is true—Humboldt apparently combined data from multiple peaks into his Chimborazo mapping project. Why? Probably because Chimbo was more famous, and his purpose was communicating broad patterns, not precise altitudes. There is some question as to whether he even collected plants at the upper elevations of Chimborazo at all. So we must be careful. Calls for rechecking came from other scientists when the first reanalysis of the Chimbo map was released, over valid concerns about the provenance of the original data, missing samples, and the like. A separate analysis of the altitudes of plants in the region had them moving upslope about half as far as originally thought, 1000 ft. (300 m) instead of 2000 ft. (600 m), although actual distance varies by species, of course. But rather than seeing this as a muddle of dubious reliability, we should understand that the process of science is messy but self-correcting—inevitable errors are discovered and rectified, facts double-checked, and conclusions constantly challenged (this particular battle went back and forth in the *Proceedings of the National Academy of Sciences*, one of the premier scientific journals in the world, for 5 years! See Suggested Reading for specifics). It would be a mistake to see this as problematic in reaching the broad conclusion: climate change is causing biology to migrate uphill. That fact is clearly evident here

FOLLOWING Humboldt's map of Chimborazo (the *Tableau Physique* in the French version) is spectacular in both scale and scope. By displaying the entire mountain simultaneously, it visualizes the orderly and predictable way in which life changes with altitude (the side panels are information on precipitation and temperature at different elevations). It also places humans on the map—but as just another part of a larger world. This visual organization inspired the first biogeographers and enabled a new, global way of thinking. Maps prior were typically literal representations of space, records of political boundaries, or for navigation, not vehicles of scientific questioning and hypothesizing. By viewing the natural world deliberately at its natural scale, and with scientific data as part and parcel of the whole, Humboldt demonstrated the use of maps for new understandings of natural processes much bigger than what typical humans perceived.

ÉCHELLE en MÈTRES	RÉFRACTION à 50° de hauteur exprimée en Secondes et pour la 1ère vers pour la Température 0°.	DISTANCE à laquelle les Montagnes sont visibles sur mer, en faisant abstraction de la réfraction.	HAUTEURS MESURÉES en différentes parties DU GLOBE.	PHÉNOMÈNES ÉLECTRIQUES Selon la hauteur des Couches.	CULTURE DU SOL selon son élévation au-dessous du Niveau de la Mer.	DÉCROISSEMENT de la Gravitation exprimé par les Oscillations d'un même Pendule dans le Vuide.	ASPECT du Ciel azuré exprimé en degrés de Cyanomètre.	DÉCROISSEMENT de l'Humidité de l'Air exprimé en Degrés de l'Hygromètre de Saussure.	PRESSION de l'Air Atmosphérique exprimée en haut Baromédrique.	ÉCHELLE en TOISES

GÉOGRAPHIE DES PL[ANTES]

Tableau physique

Dressé d'après des Observations & des Mesures...

jusqu'au 10.° de latitude au...

ALEXANDRE DE H[UMBOLDT]

Esquissé et rédigé par M. de Humboldt, dessiné par Schönberg...

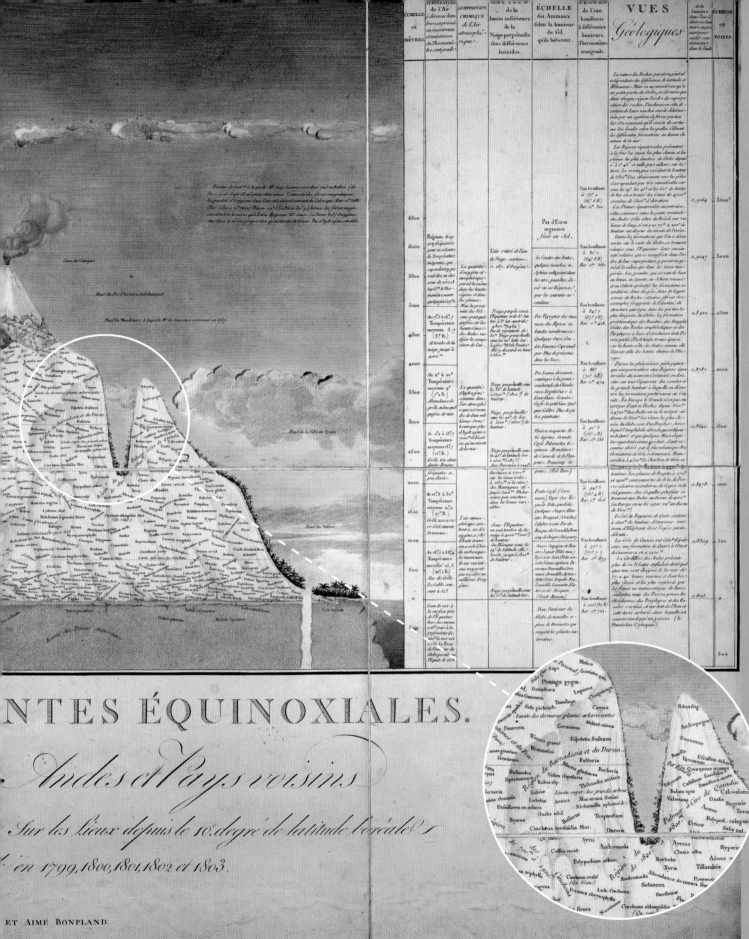

and elsewhere. Rather, this scientific "conversation" shows us both the limits and the amazing benefits of historical data as part of our evolving view of the world.

The goal of this book is to take the fundamental science of our day, as it applies to the natural and human world, and explore the basic systems that make life on Earth possible. To make it relatable and understandable, the text frequently utilizes case studies and localized stories. This brings the science from the realm of the abstract to the realm of lived experience. To connect those local stories to the global system they exemplify, though, requires a bit of Humboldt: historical charts for the context of time and modern technology for the context of space—via satellites, global models, and art. The maps and illustrations you will find here are an equal partner with the written information, the tool by which we scale that lived experience to a global framework—which is, of course, the proper scale to study our world and how it is changing. It starts with the atmosphere, probably the most familiar of climate-related phenomena, and then moves to the oceanic and terrestrial dimensions. After that, we will explore the urban setting, a uniquely human habitat created, in many ways, in our own image. The book concludes with an exploration of life's biodiversity and the ways in which it is described and maintained as well as how it functions. The themes are those of scale and variability—how we experience the world versus how it works, and how variability (human caused or otherwise) is accommodated, endured, or forced into submission. There is much more, of course, that is omitted (all apologies to the wondrous benthic ocean ecosystems, for example!), a sad necessity of a single book and a big globe. As you move through the imagery and text, occasionally take the time to close your eyes and situate yourself in the larger system of the World, capital "W," and feel your place in the broader story of the earth system.

Historical work continues to contribute to new insights. Recent research replicating Humboldt's original survey provides visceral evidence of climate change in the migration of species uphill. While the original *Tableau* was a conglomeration of multiple mountains (summarized and adapted on the left)—and not necessarily up to the rigorous standards of modern scientific mapping—it still provides an invaluable reference point. Resurveys and additional studies on those neighboring mountains and challenging scientific review have supported the general idea that Humboldt's mountains have changed dramatically because of climate warming (seen on the right).

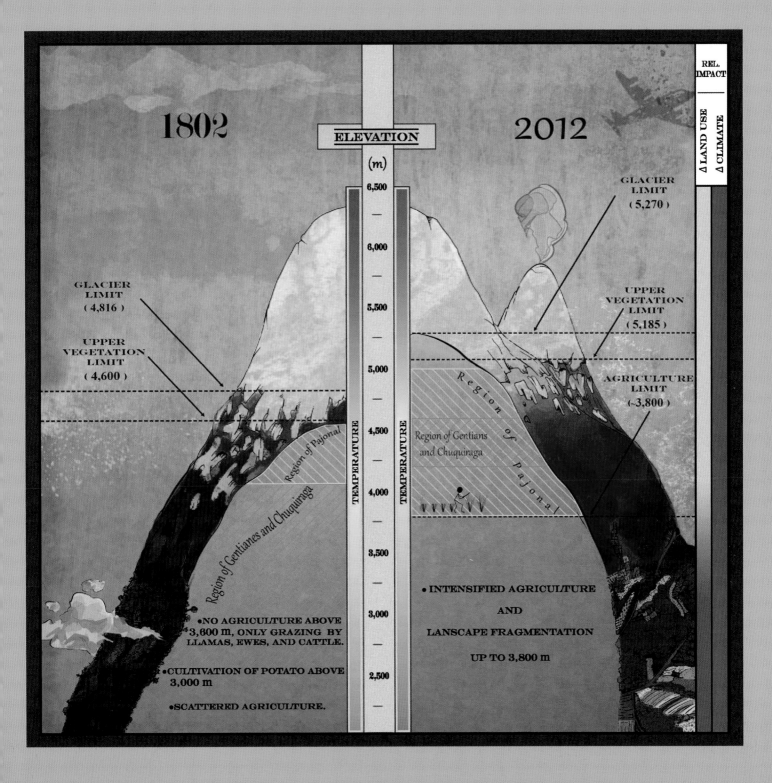

1802

2012

ELEVATION

(m)

GLACIER
LIMIT
(4,816)

UPPER
VEGETATION
LIMIT
(4,600)

GLACIER
LIMIT
(5,270)

UPPER
VEGETATION
LIMIT
(5,185)

AGRICULTURE
LIMIT
(~3,800)

Region of Pajonal

Region of Pajonal

Region of Gentians
and Chuquiraga

Region of Gentianes and Chuquiraga

- NO AGRICULTURE ABOVE
3,600 m, ONLY GRAZING BY
LLAMAS, EWES, AND CATTLE.

- CULTIVATION OF POTATO ABOVE
3,000 m

- SCATTERED AGRICULTURE.

- INTENSIFIED AGRICULTURE

AND

LANSCAPE FRAGMENTATION

UP TO 3,800 m

6,500

6,000

5,500

5,000

4,500

4,000

3,500

3,000

2,500

TEMPERATURE

TEMPERATURE

REL.
IMPACT

Δ LAND USE   Δ CLIMATE

ATMOSPHERE

ALWAYS IN MOTION

One of your most intimate relationships is with the atmosphere. It determines what you wear each day, how long your commute takes, the price of your meals, and the safety of your home (among many other aspects of life). Weather is the first thing people think about in the morning and the most common (if bland) conversation opener. "Beautiful day, isn't it?" "Some rain we're having, huh?" That intimate familiarity is both a blessing and a curse. The atmosphere can fade into the background, and many assume their experience with the weather is representative of all—a mistake of scale, a mistake of perception. This type of misperception led to the infamous February 2015 incident in which US Senator James Inhofe brought a snowball to a speech to demonstrate the implausibility of global warming. The weather *where I am* is unseasonably cold, therefore *global* warming is a hoax. That is the logical mistake. Wherever you are is only a small part of the whole, and not necessarily representative of the entire system. That example is doubly unfortunate; not only does it represent a fundamental misunderstanding of global environmental processes at the highest levels of US governance, but 2015 was globally the hottest year ever recorded at the time. (Since then, we have broken the record three more times: 2016, 2019, and 2020.) So let's think about the basics. It's worth it.

How much do you really know about the atmosphere at the scale it functions—globally? We can start with something basic: Where does your weather come from? And then the commonality angle: Where does your air go?

THE GLOBAL POOL

How knowledgeable are you about the air you breathe? Many can get to the primary school level—it's about 78 percent nitrogen and 21 percent oxygen, with the remainder a cocktail of carbon dioxide, argon, and other molecules. There are even some less savory components, though they tend to vary considerably with time and space—carbon monoxide (at a numerical concentration around 0.000005 percent), sulfur dioxide (around 0.0001 percent), and ammonia (around 0.0000003 percent). Human pollution sources can drive those harmful products up, of course, as can phenomena like volcanic eruptions. Take a deep breath (in an unpolluted setting!). Like fish in a carefully balanced aquarium, we are submerged in this airy medium, day in and day out, for better or worse.

The atmosphere is a global pool about 50 mi. (80 km) thick (at least the lower three layers) and we are all swimming in it—or rather, walking along its bottom. Birds approximate swimming, at least temporarily. In many ways, water is not a bad analogy; the atmosphere swirls around peaks, ponds in valleys, and slushes out through mountain passes in a way very similar to water currents. Consequently, it is a fairly well-mixed system, diluting small amounts of pollutants to our benefit but spreading the impact of major issues globally. Much of this is invisible, so tracing the ebb and flow of the atmosphere requires a combination of measurements in specific spatial locations, sensors capable of seeing electromagnetic radiation in

HOW MUCH
DO YOU REALLY
KNOW ABOUT
THE ATMOSPHERE
AT THE SCALE
IT FUNCTIONS—
GLOBALLY?

wavelengths our eyes can't see, and computer models to interpolate those measurements across space.

The results are astounding, beautiful, and appalling in equal measure. A map of particulate matter in the atmosphere (microscopic particles carried along in the currents) created by NASA shows the spread of that fluid drama around the world. Beautiful bands of high salt concentrations are spun up throughout the southern hemisphere by storms circling the Roaring Forties, Furious Fifties, and Screaming Sixties—all references to latitudes that were a source of terror and excitement during the Age of Sail. Those storms reach incredible magnitudes both in intensity and spatial extent because of the lack of land to break them up, and you can see the long tendrils of salt lofted off the waves and spun, like a web, across the Southern Ocean.

But there are less romantic components of particulate maps as well. Combustion material, often called black carbon because it is essentially charcoal in microscopic form, is also carried along in swirling plumes across the oceans. The blast of acrid smoke from wildfires in the western United States and Canada plus agricultural fires in Africa is clearly observable, but so is the low-level and more widespread noxious leaking of carbon pollution from human centers like the eastern US and China. Once entrained in the jet stream, that carbon can loop and twist its way across the Atlantic or Pacific, spreading black carbon to remote locations where it rains out on Earth's surface and causes a variety of impacts, such as the acceleration of glacial melt (because it darkens the snow's surface). The other side of the world is not actually that far away when we're talking about the atmosphere.

This means that no one, no matter in what country or environment, is isolated from anyone else. This is true for time as well as space. What we put into the air doesn't always disappear with the seasons; the world has a long memory. Sir James Hopwood Jeans, an eminent English physicist and astronomer, illustrated it beautifully in 1940:

> If we assume that the last breath of, say, Julius Caesar has by now become thoroughly scattered through the atmosphere, then the chances are that each of us inhales one molecule of it with every breath we take…
> —Sir James Hopwood Jeans, *An Introduction to the Kinetic Theory of Gases*

A person's lungs hold about 2 liters of air (a bit more than any individual breath), so your lungs could plausibly have about 5 molecules from Julius Caesar's

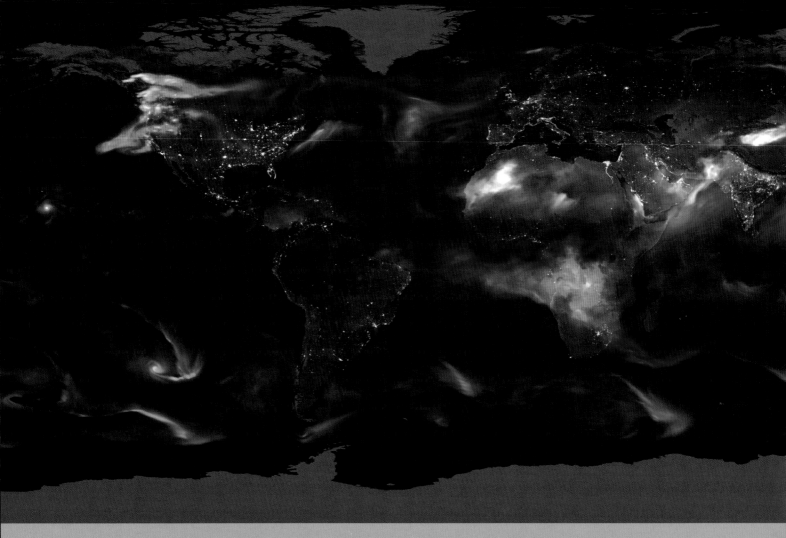

NO ONE, NO
MATTER IN WHAT
COUNTRY OR
ENVIRONMENT, IS
ISOLATED FROM
ANYONE ELSE.

THE
ATMOSPHERE
IS A GLOBAL
POOL . . . AND
WE ARE ALL
SWIMMING
IN IT.

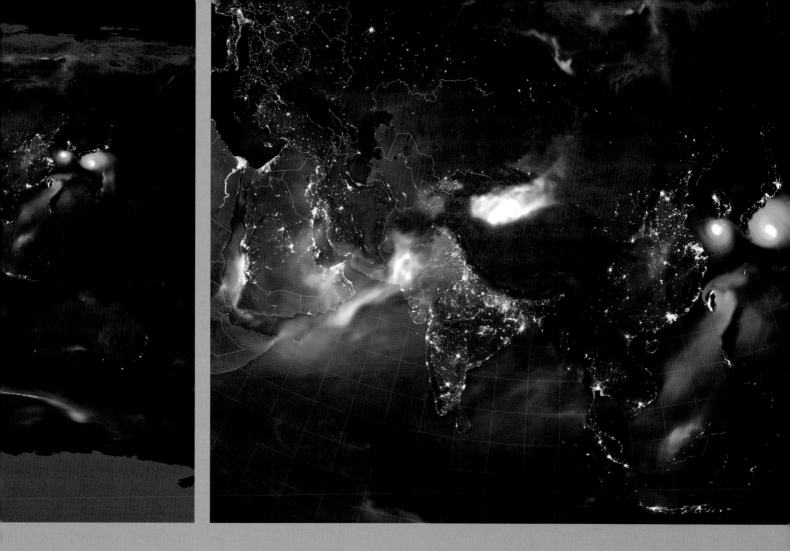

Particle Pool

Particles generated by natural and human processes spin throughout the atmosphere on global circulation paths that connect each one of us, lung to lung. These maps, outputs from a NASA Global Circulation Model platform for the day of August 23, 2018, show salt (blues), black carbon from fires and combustion engines (reds), and dust (purple) in our air. The maps illustrate both the hotspot sources of these particles, such as dust from deserts and sea salt from hurricanes, and their reach: black carbon from North American fires float across the North Atlantic Ocean, entrained on the westerlies, to land on alpine glaciers. Dust from the African Sahara fertilizes the tropical rainforests of South America with phosphorus, a necessary nutrient. A bit of Africa feeds the new world—connections across continents carried by the atmosphere.

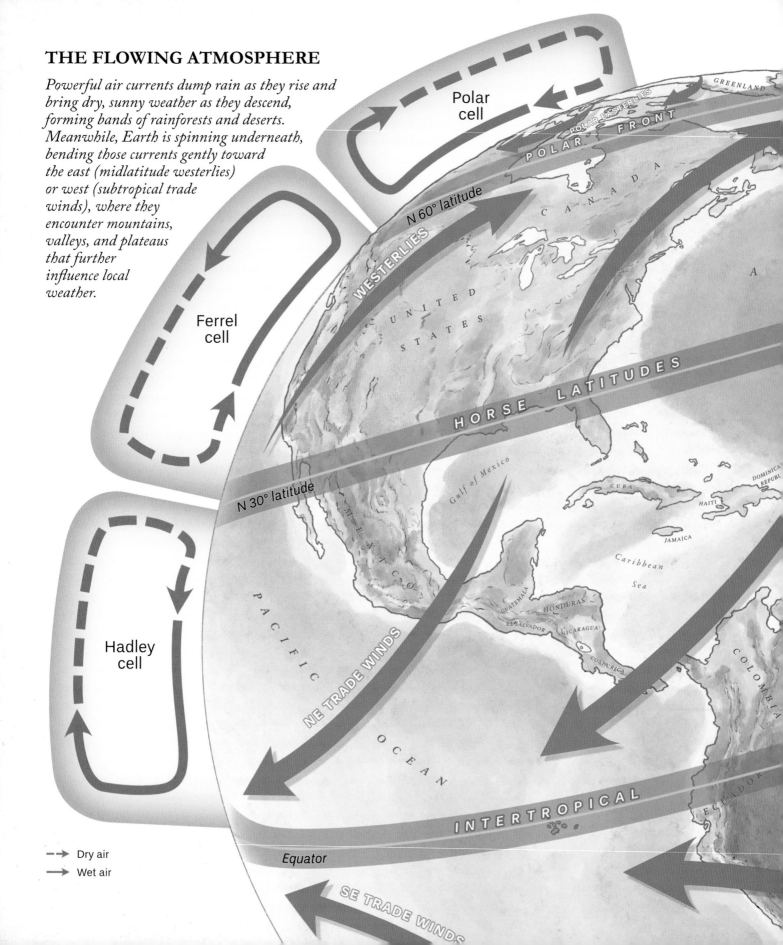

THE FLOWING ATMOSPHERE

Powerful air currents dump rain as they rise and bring dry, sunny weather as they descend, forming bands of rainforests and deserts. Meanwhile, Earth is spinning underneath, bending those currents gently toward the east (midlatitude westerlies) or west (subtropical trade winds), where they encounter mountains, valleys, and plateaus that further influence local weather.

Polar cell

Ferrel cell

Hadley cell

GREENLAND

POLAR EASTERLIES

POLAR FRONT

N 60° latitude

C A N A D A

WESTERLIES

U N I T E D   S T A T E S

A

N 30° latitude

H O R S E   L A T I T U D E S

Gulf of Mexico

CUBA

DOMINICAN REPUBLIC

HAITI

JAMAICA

Caribbean Sea

M E X I C O

NE TRADE WINDS

GUATEMALA

HONDURAS

EL SALVADOR

NICARAGUA

COSTA RICA

COLOMBIA

P A C I F I C   O C E A N

I N T E R T R O P I C A L

ECUADOR

Equator

SE TRADE WINDS

- -→ Dry air
- → Wet air

As winds rise in elevation to pass over obstacles, air pressure decreases. And as air pressure declines, temperature drops. Cooler air can't hold the moisture that warmer air can, and out it comes—as rain or snow blanketing the windward side of mountains.

WESTERLIES

A T L A N T I C

NE TRADE WINDS

O C E A N

MOROCCO
WESTERN SAHARA
MAURITANIA
SENEGAL
THE GAMBIA
GUINEA-BISSAU
GUINEA
SIERRA LEONE
LIBERIA
CAPE VERDE

Earth's spin

Even at the finest scales, air is a dynamic, playful thing. Albatrosses can fly hundreds of miles a day without flapping their wings—by dipping into calm troughs between waves, then popping over the crests to catch higher-speed winds. The birds then soar up about 30 ft. (10 m), turn over, and swoop back down in a tacking, circular pattern called "dynamic soaring."

Wandering albatross
(Diomedea exulans)

VENEZUELA
GUYANA
SURINAME
FRENCH GUIANA

C O N V E R G E N C E   Z O N E   (I C Z)

B R A Z I L

last breath (assuming those molecules are still there, and have not been absorbed into the oceans, metabolized, or met some other fate).

There is no getting away from the fact that the atmosphere is a common blessing, but also a common responsibility. Our breaths are shared with everyone else, past, present, and future.

GLOBAL CHANGE AND ATMOSPHERIC DATA

The clear issue of our times is carbon (primarily carbon dioxide, CO_2) in the atmosphere. Carbon dioxide is a natural and necessary atmospheric component, at historical proportions of of about 0.027 percent, and now around 0.042 percent, depending on the season. This small fraction is needed to make Earth habitable: our climate is very sensitive to additions, or removals, of CO_2.

The idea of climate change as a result of carbon emissions is not new; scientists have long appreciated the significance of CO_2 to global climate. In the early 1800s, French mathematician and physicist Jean Baptiste Joseph Fourier realized the atmosphere must be retaining heat in some fashion. He compared Earth's atmosphere to a glass box around the planet, which became the basis of our "greenhouse" analogy. Fourier's research inspired John Tyndall, an accomplished man in many scientific fields, to proclaim, "As a dam built across a river causes a local deepening of the stream, so our atmosphere, thrown as a barrier across the terrestrial rays, produces a local heightening of the temperature at the Earth's surface." His work with CO_2 and water vapor conclusively demonstrated the heat-retaining abilities of both. Now we know how it works: CO_2 has a molecular structure that can absorb longwave radiation coming off Earth's surface (which otherwise could sail out into space), warming and exciting the molecule. It eventually reradiates that heat in a random direction, and some of the heat goes back to the ground, rather than space. Thus, heat accumulates. Tyndall didn't know the mechanism at the time, but his work inspired Swedish Nobel Prize winner Svante Arrhenius, who first tied those greenhouse gases to a changing climate and attempted to calculate the actual work accomplished by that CO_2 (he estimated an increase of approximately 10°F [5.5°C] with a doubling of CO_2; the actual warming potential is probably about half that— but it's still an impressive calculation for the time) and in a second inspired leap, connected it to human activity. That was in 1896.

By the middle of the 20th century, we knew enough to take this potential seriously. The challenge was measuring CO_2 in such a way that it was as pure a baseline

PREVIOUS
The air we breathe is global and dynamic, spreading vertically and horizontally around the planet, driven by heating near the equator and the rotation of the earth. Like insects on the bottom of a fast-moving creek feeling the currents, our weather systems swirl and streak overhead, driven by this global-scale system that brings warmth and rain in their seasons to some, cold and dry to others.

OPPOSITE
Each year, carbon dioxide rises and falls a few parts per million (ppm) with the seasons. But it never quite falls as far as the year prior, and so atmospheric carbon dioxide continues to accumulate and rise. In black, the up and down with the seasons; in red, the increase in CO2 after removing the annual fluctuations.

as possible, representative of the global scale. Charles (Dave) Keeling, at California Institute of Technology, took up that challenge. At the time it was no trivial task. The precise instrumentation required a chemist who was a wizard with technicalities. The process involved measuring the precise equilibria between water, limestone, and atmospheric CO_2. Keeling was working in Pasadena, California, and found he couldn't see the air for the industry—local sources were swamping both baseline and natural variations in CO_2. Keeling spent a considerable amount of time tracking down the causes of the diurnal (day-night) fluctuations in CO_2, hauling his apparatus from California to rainy Washington to more-rainy Amazonia. The answer, that CO_2 declines in the daytime because of high rates of photosynthesis and increases at night because of plant respiration (with no concurrent photosynthesis), was but the beginning of perhaps the most significant long-term atmospheric scientific project to date.

Keeling's work got him an invitation to work on a global network of CO_2 sensors to monitor atmospheric carbon, stretching all the way to Antarctica. While all the sensors generated interesting local data, the one that really stuck was on an extinct volcano in Hawaii. Mauna Loa was not the best place for Keeling's CO_2 measurements because the weather is nice; Hawaii's location in the middle of the Pacific

THE IDEA OF CLIMATE CHANGE AS A RESULT OF CARBON EMISSIONS IS NOT NEW.

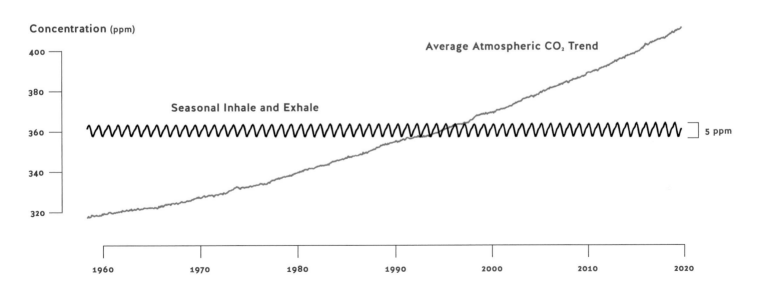

Ocean, close(ish) to the equator, and with little but water upwind means it gets as well-mixed a stew as possible, with no nearby power plants mucking up the data.

In March 1958, the first atmospheric carbon dioxide reading spooled out of the infrared gas analyzer into Keeling's lap: 313 parts per million or ppm (for reference, we were at around 419 ppm in April 2021). Over the next months, Keeling watched the CO_2 counter rise to a maximum of 317.5 ppm in May—but then fall, reaching its lowest count in October, when the measurement was 312.7 ppm. Over the next six months, he saw the inevitable exhale as the counter rose to a new peak in May 1959, at 318.3 ppm. Keeling was the first to observe the Earth breathe.

The amount that is measured in the air fluctuates depending on the time of year. Lower levels dominate during the northern hemisphere summer; the area is mostly terrestrial with lots of plants. Higher levels are measured during the northern hemisphere winter and early spring because of the decline in photosynthesis due to the cold (and relative lack of photosynthesis in the southern hemisphere to pick up the slack).

The record has been maintained ever since. Carbon dioxide continues to rise, thickening the blanket around the world by trapping more and more heat. Today, the average additional heating caused by human CO_2 emissions (i.e., the amount of solar heat trapped that would not have been otherwise) is about 1 or 2 watts per square meter (a square meter is about 20 percent bigger than a square yard). That may not sound like much—around 3 or 4 non-LED Christmas lights. But imagine those lights on, all the time, day and night, everywhere on the entire surface of the planet, the surplus heat slowly accumulating. The inexorable rise of CO_2 from fossil fuel burning has been inevitably followed by warming, just as Arrhenius anticipated over a century ago.

FROM LOCAL WEATHER TO ATMOSPHERIC BEHAVIOR

Our slowly warming pool, shared by us all, illustrates that there's nothing truly independent about our world. Pollution from industry in North America alters Asian temperatures, and vice versa. Every time we zoom out in scale, we find a new connection between places—and not just in climate, but also in our daily weather. Take major storms and hurricanes, for example. They are clearly an immediate and real threat to people in their path, or even near the path. But until recently, our ability to forecast hurricanes was incredibly poor because we couldn't capture the scale we needed to understand their behavior, and our mapping was too deficient in data. In the modern era, with geostationary satellites and a dense network of instruments worldwide, it is hard to remember how difficult it was to see over the horizon only a few decades ago.

Early work on storms and hurricanes had to piece together atmospheric circulation that spanned thousands of miles from only a few observations, often collected opportunistically, like records on wind speeds from ships that just happened to be crossing the Atlantic at the right time. These measurements would be made on the deck of a heaving ship, with rudimentary instruments coated in slime and seawater and manned by exhausted seamen. The quality of such measurements could be dubious, but they were indispensable. Records would be brought together later, with early meteorologists trying to find swirls of organized storms peeking through the sparse chaos of wind direction, speed, pressure, and humidity measurements. It was an art. When done wrong, the 50 percent chance of sun would turn out to be rain. When done right, a picture of a hurricane would appear.

One of the first examples of this mash-up of meteorology and broadscale spatial-scientific-artistic intuition was published as the first map in the first issue of *National Geographic*, in 1888, as part of a story by Edward Hayden, depicting North America's "Great White Hurricane of 1888." (Note—this was not actually a hurricane in the technical sense but did reach hurricane-force winds). Hayden's map was stitched together from a collection of United States Signal Service (USSS) weather stations (170 in total across the US, observing three times a day), volunteer weather observers (up to 2000 in the US depending on the time of day and who was reporting), and measurements from a variety of ships stretching from coastal Florida halfway to Iceland. The magnitude and direction of wind are denoted with arrows and temperature, running from the toasty red of Florida to the frigid blue of interior Canada. This was a map people wanted to see, not to direct evacuations or give timely warnings—the map was published six months after the storm!—but because of the impact and what it might reveal about why that storm was so bad.

The storm had hammered New England in mid-March, killing 400 people and dropping more than 6.5 ft. (2 m) of snow. Drifts reached nearly 40 ft. (12 m) high in some areas. Roads were impassable; livestock and horses died by the thousands. In the Berkshires, a rural part of the region, people were stuck in their homes for two weeks. That rudimentary map of Hayden's, so basic today, explains why. The

OUR SLOWLY WARMING POOL, SHARED BY ALL OF US, ILLUSTRATES THAT THERE'S NOTHING TRULY INDEPENDENT ABOUT OUR WORLD.

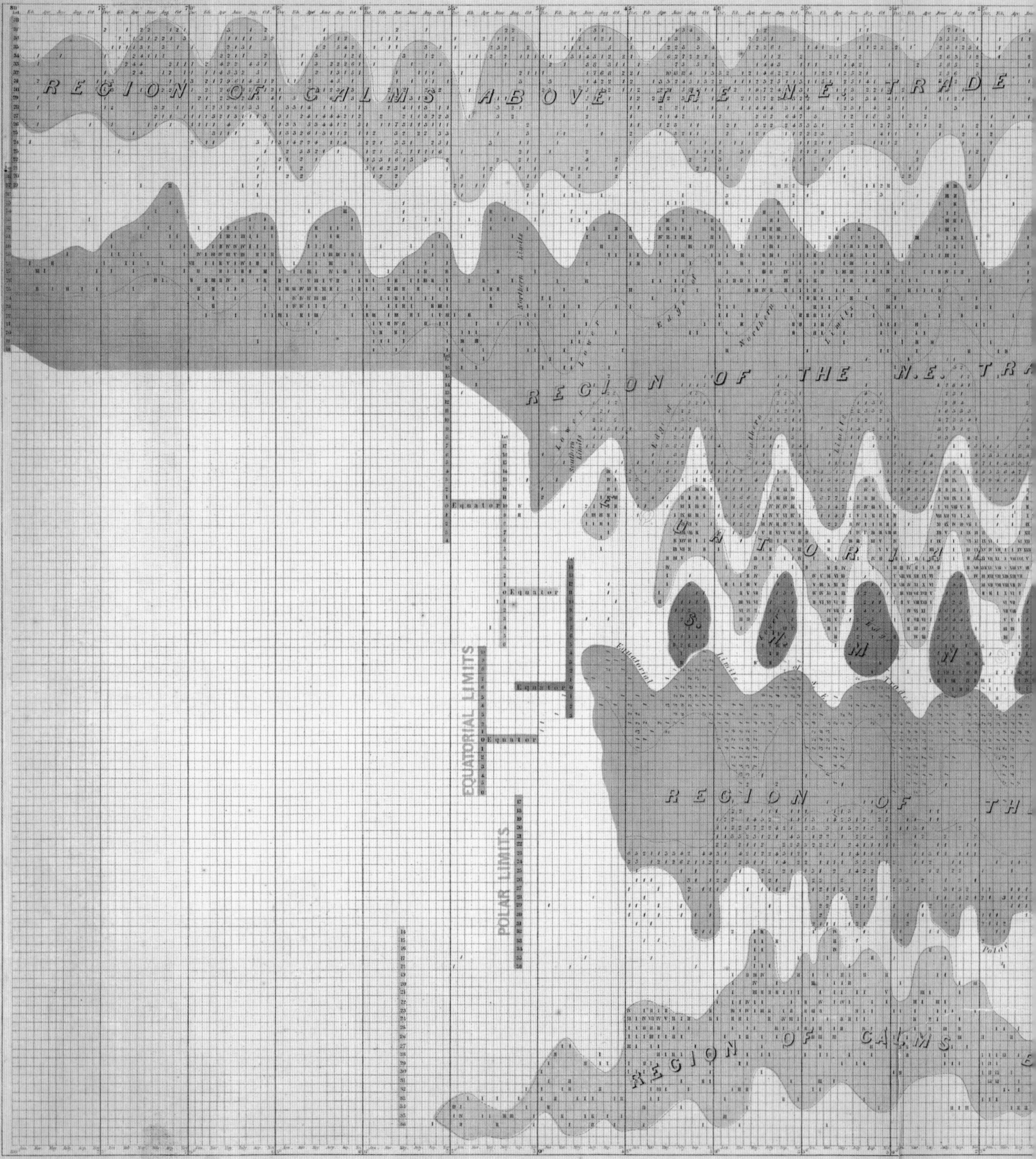

REGION OF CALMS ABOVE THE N.E. TRADE

REGION OF THE N.E. TRA

EQUATORIAL LIMITS

POLAR LIMITS

o Equator

o Equator

o Equator

o Equator

Northern Limits

Lower Southern Limits

Northern Limits

REGION OF THE

REGION OF CALMS

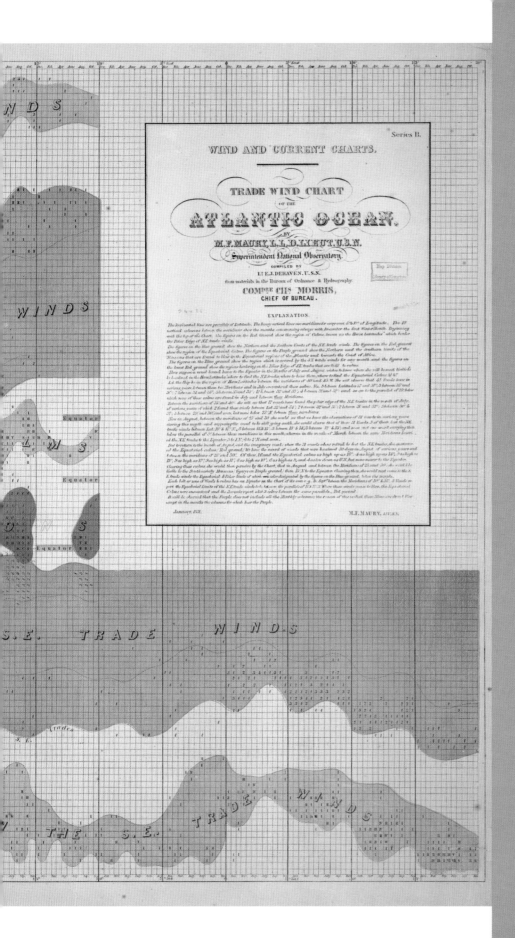

Published by Matthew Fontaine Maury in 1851, these early records of winds chart the annual fluctuation of the all-important trade winds and the regions of calms (also known as the horse latitudes), which are disastrous for a sailing ship. It was a major achievement of climatological data synthesis; despite its appearance, it is actually an amazing ocean-scale accomplishment of environmental charting. The horizontal lines represent latitude, the vertical lines are months of the year for different longitudinal regions, west to east. The numbers are the observations in support of this data map, one of the earliest of its kind. The annual procession of the trade winds north and south, a result of the seasons, is reflected in these data—waves of blue upward as winds shifted north in summer, downward as they returned south in winter.

A few short years later, the Indian Ocean also had a chart of this nature; the Pacific Ocean was still too unknown to map in this fashion.

"hurricane" system of the storm was spread out in a large trough, a deep low-pressure system that pulled cold air down and east into warmer, wetter air; this caused an immense amount of snow. The trough was very stable, remaining over southern New England for 48 hours. The sharp edges of the trough, evidenced on Hayden's map, show areas of intense wind that buried houses in drifts of snow—wind flows from high to low pressure areas, and just like the difference between a gentle stream and a waterfall, the horizontal distance between the high and low points matters a lot: a fast, steep drop can be a violent affair. It is clear enough in retrospect. But at the time, those weather stations were not enough to be predictive. Focus on too small an area, and the system doesn't look so intimidating. "Fresh to brisk easterly winds, with rain, will prevail tonight followed on Monday by colder, brisk westerly winds and fair weather throughout the Atlantic states," was forecast for March 11 and 12 from the USSS. Part of their enormous error was due to the lack of scientific knowledge about atmospheric behavior and a lack of computing power to forecast the future. But that was only a part. The bigger issue was the lack of measurements that matched the scale of the problem.

Today, we can map and accurately model wind and weather at a much broader scale. And what we see reminds us exactly how large things like hurricanes actually are. When Hurricane Sandy struck New Jersey in the fall of 2012, the direct ripple effects stretched all the way to the Great Plains. The indirect effects—the changes in storm tracks which both shape and are shaped by these immense hurricane systems—could be tracked around the world. Today, modern models and the immense amounts of data on wind speed, pressures, and temperatures from both ground and space-based instruments allow for highly accurate forecasts several days before a storm hits—truly a lifetime if you successfully evacuate. They also illustrate that we are all on the edges of hurricanes; again, from an atmospheric standpoint, no one is truly separate from anyone else.

TEMPERATURE AND PRECIPITATION

The most immediate and familiar aspects of the atmosphere—and the most obviously sensitive to human-caused change—are temperature and precipitation. Our very lives are tied up with warm air and rains, from immediate existential threats such as flooding to prosaic but also life-threatening issues such as drought and food production. Indeed, as soon as our ancestors in the Middle East, East Asia, and North America tied our lives up with intensive agriculture, local temperature and precipitation became the subjects of vital interest. Once humans started globalizing our

This is the original weather chart of the onset of the Great White Hurricane, which occurred March 11–14, 1888. It was constructed by Edward Hayden from observations and published late that same year. Colors are isotherms of temperature, arrows indicate wind direction and strength, and numbers indicate barometric pressure.

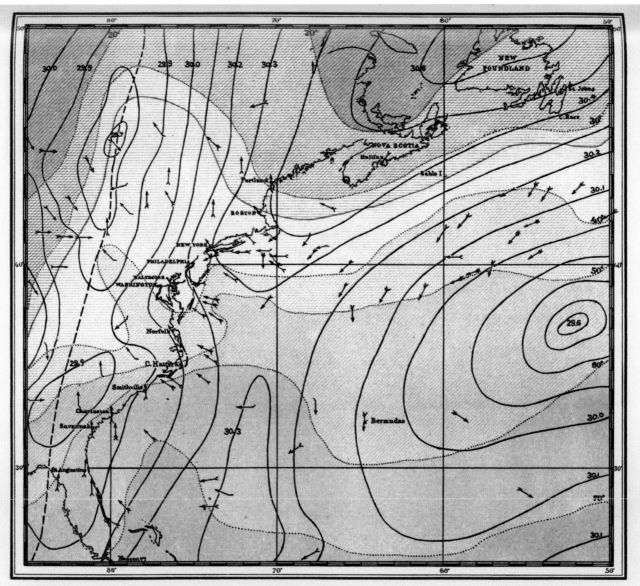

Barometer.—Isobars in full black lines for each tenth of an inch, reduced pressure. The trough of low barometer is shown by a line of dashes.

Temperature.—Isotherms in dotted black lines for each ten degrees Fahr. Temperatures below freezing (32° F.) in shades of blue, and above freezing in red.

Wind.—The small black arrows fly with the wind at the position where each is plotted. The force of wind is indicated in a general way by the number of feathers on the arrows, according to the scale given in the following table:

PLOTTED ON CHART.	FORCE, BY SCALES IN PRACTICAL USE.					POUNDS PER SQUARE FOOT.	MILES PER HOUR.	KILOMETERS PER HOUR.	METERS PER SECOND.
	0 — 12	0 — 10	0 — 8	0 — 7	0 — 6				
○ Calm.	0	0	0	0	0	0.	0.	0.	0.
→ 1	1 — 2	1 — 2	1	1 — 2	1	0. — .40	0. — 9.	0. — 14.4	0. — 4.
→ 2	3 — 4	3 — 4	2	3 — 4	2	0.41 — 2.53	9.1 — 22.5	14.5 — 36.2	4.1 — 10.1
→ 3	5 — 7	5 — 6	3 — 4	5	3	2.54 — 8.20	22.6 — 40.5	36.3 — 65.2	10.2 — 18.1
→ 4	8 — 10	7 — 8	5 — 6	6	4 — 5	8.21 — 22.90	40.6 — 67.5	65.3 — 108.7	18.2 — 30.1
→ 5	11 — 12	9 — 10	7 — 8	7	6	22.91 and over.	67.6 and over.	108.8 and over.	30.2 and over.

It will be noticed that the Beaufort scale (0–12), in general use at sea, has been converted into the international scale (0–10) for the sake of clearness in plotting data on the chart. The absence of arrows over large areas indicates absence of simultaneous data; at sea, however, this has been partly compensated for in the construction of the chart by information obtained from journals and special storm reports of vessels in the vicinity.

Hurricane Sandy
and its extreme winds,
estimated from a NASA
high-resolution weather
model of the 2012
mega-storm.

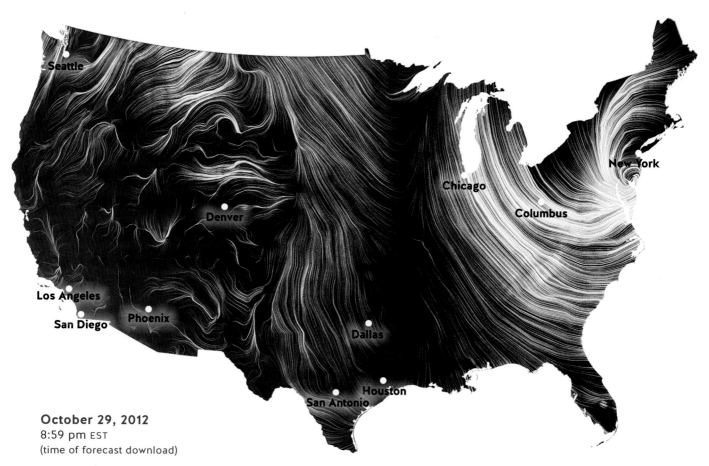

Seattle

Denver

Chicago

Columbus

New York

Los Angeles

San Diego

Phoenix

Dallas

Houston

San Antonio

October 29, 2012
8:59 pm EST
(time of forecast download)

top speed: **45.1 mph**
average: **9.4 mph**

The effects of Sandy were felt across the
continent, either directly, in the form of winds
or wind paths, or indirectly, by the way in which
the hurricane blocked or directed other weather
systems. This map traces wind directions and
magnitudes right at landfall, clearly illustrating how
hurricanes influence far more than their immediate
vicinity.

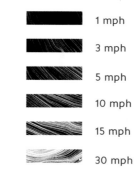

1 mph

3 mph

5 mph

10 mph

15 mph

30 mph

Hemispheric Hurricanes

FOLLOWING SPREAD This unique perspective, a global view centered over the South Pole, shows all recorded hurricane tracks since the mid-1800s and reveals that hurricanes are, surprisingly, a northern hemisphere problem (the dashed line is the equator; the green and blue tracks are individual hurricanes and brighter areas indicate overlapping tracks). Why are there fewer hurricanes in the southern hemisphere? We have to think about the atmosphere and oceans together. Hurricanes thrive in warm water. Oceans in the southern hemisphere are considerably colder than their northern counterparts because of their massive size and the ice on and around Antarctica. Another major difference is the greater wind shear, or the force of incoming winds, on potential hurricanes in the southern hemisphere. If wind shear is too high, hurricanes cannot organize into the tight spirals that power their heat engines. But how climate change will influence southern hemisphere hurricane formation is still an open question.

TODAY, WE CAN MAP AND ACCURATELY MODEL WIND AND WEATHER AT A MUCH BROADER SCALE.

HURRICANES

& Tropical Storms | Locations & Intensities since 1851

AUSTRALIA

ANTARCTICA

SE ASIA

AFRICA

John Nelson | uxblog.idvsolutions.com
IDV Solutions | idvsolutions.com

NOAA International Best Track Archive | ncdc.noaa.gov
NASA Visible Earth | visibleearth.nasa.gov

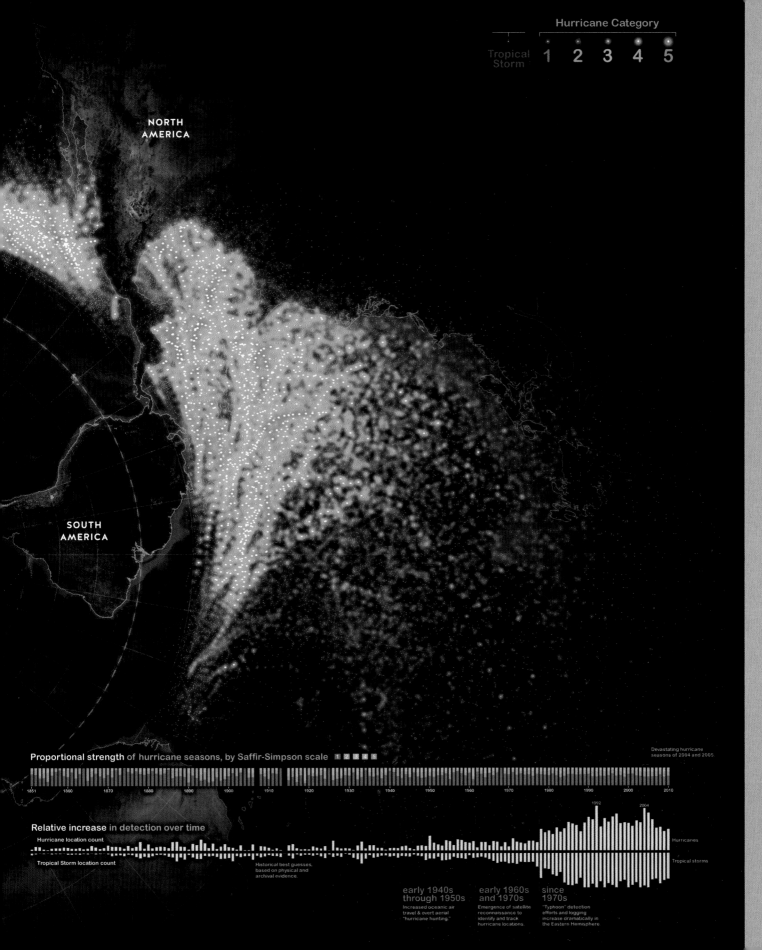

Hurricane Category

Tropical Storm   1   2   3   4   5

NORTH
AMERICA

SOUTH
AMERICA

Proportional strength of hurricane seasons, by Saffir-Simpson scale 1 2 3 4 5

Devastating hurricane
seasons of 2004 and 2005

1851   1860   1870   1880   1890   1900   1910   1920   1930   1940   1950   1960   1970   1980   1990   2000   2010

1992   2004

Relative increase in detection over time

Hurricane location count

Historical best guesses,
based on physical and
archival evidence.

Tropical Storm location count

Hurricanes

Tropical storms

**early 1940s
through 1950s**
Increased oceanic air
travel & overt aerial
"hurricane hunting."

**early 1960s
and 1970s**
Emergence of satellite
reconnaissance to
identify and track
hurricane locations.

**since
1970s**
"Typhoon" detection
efforts and logging
increase dramatically in
the Eastern Hemisphere.

supply chains and trading for far-flung commodities, mapping differences in temperature and precipitation—not just gathering data at a place but truly visualizing *across space*—became necessary as an empire-management tool. And so we have some of the earliest scientific studies utilizing maps and charts at broad scales focusing on the relatively simple effects of temperature and precipitation changing through space and time.

Temperature and its steady increase

Temperature is probably the most familiar aspect of climate, and average (or mean) temperature is one of the most familiar aspects of climate to most people. The average of any given location is a fundamental descriptor of climate. That's why people are fairly confident that Honolulu will be warmer than Milwaukee in the winter, at least at sea level.

Spatial variation in average temperatures was the gateway drug for thinking about natural systems globally—and we're back to Humboldt for this story. Humboldt realized that climate (and particularly temperature) did not necessarily depend on how far north or south one was, at least not completely. Annual temperatures, or as he put it, annual inputs of heat, varied as a function of distance to the ocean, mountains, and a variety of other factors. He came up with the concept of the isoline (or isotherm) in 1816, which was both visionary and simple: Humboldt had the idea of taking average temperatures from weather stations around the globe and drawing lines on a map that connected sites with the same annual average ("iso-" is a Greek prefix meaning "equal"). The charting was done in the abstract, however, not taking that final step toward the real world.

The first to publish an isothermal map was not Humboldt himself but rather William Channing Woodbridge, an American geography teacher of deaf students. Woodbridge followed in the footsteps of his Yale-educated father, also William Woodbridge, who was a crusader for educational reform and women's education in New England. William Channing Woodbridge, who suffered from chronic sickness similar to tuberculosis, traveled to Europe to meet with several geographical idols, including Humboldt. Using Humboldt's data, Woodbridge generated a map that spanned much of the world in 1823—for teaching purposes rather than research. It was featured in his textbook, *Woodbridge's School Atlas to Accompany Woodbridge's Rudiments of Geography*. While he certainly knew the map was innovative and novel as a teaching tool, it is unclear whether he grasped the significance of the new perspective.

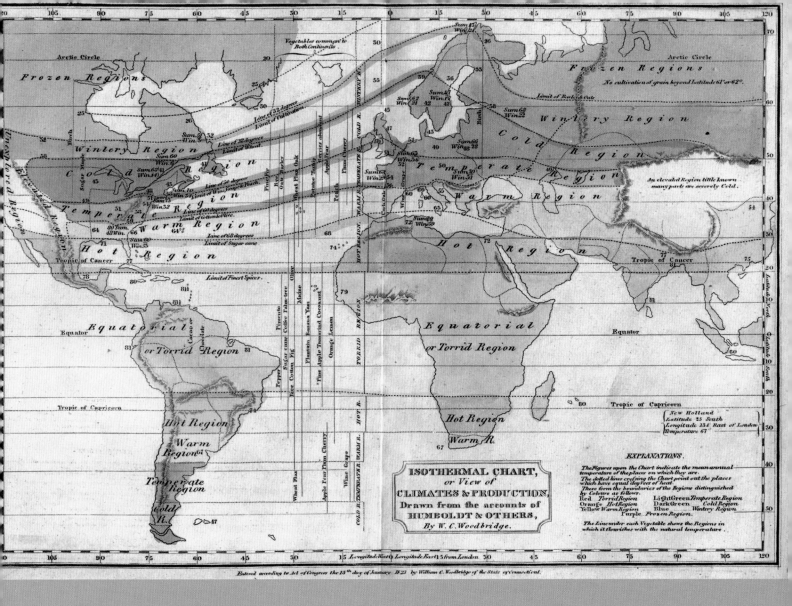

Woodbridge's Thermal Bands

Woodbridge pioneered the first attempt to map the global temperature environment on a single map, showing how various regions are connected by the common climatic system. This map, from an edition published in 1830, is from *Woodbridge's School Atlas to Accompany Woodbridge's Rudiments of Geography*, a book for school children. Temperatures run from the "frozen" blues of the arctic to the "hot" and "torrid" oranges and reds of the subtropical deserts and tropical equatorial regions.

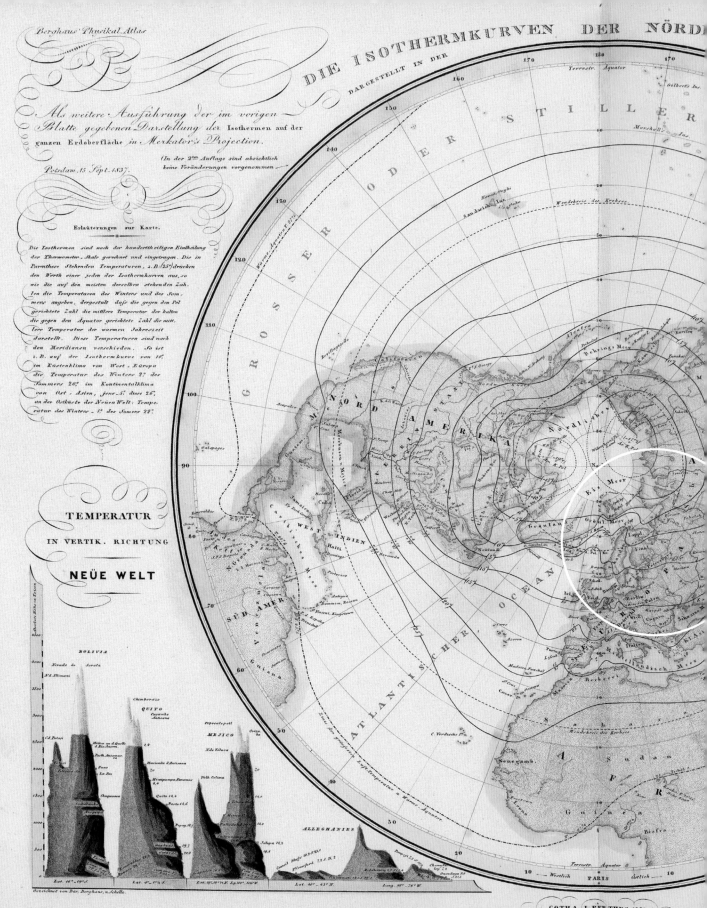

DARGESTELLT IN DER

Als weitere Ausführung der im vorigen
Blatte gegebenen Darstellung der Isothermen auf der
ganzen Erdoberfläche in Merkator's Projection.

(In der 2ten Auflage sind absichtlich
keine Veränderungen vorgenommen

Potsdam, 15 Sept. 1837.

Erläuterungen zur Karte.

Die Isothermen sind nach der hunderttheiligen Eintheilung
der Thermometer-Skale gerechnet und eingetragen. Die in
Parenthese stehenden Temperaturen, z.B.(25°)drücken
den Werth einer jeden der Isothermkurven aus, so
wie die auf den meisten derselben stehenden Zah-
len die Temperaturen des Winters und des Sommers angeben, dergestalt dass die gegen den Pol
gerichtete Zahl die mittlere Temperatur des kalten
die gegen den Äquator gerichtete Zahl die mittlere Temperatur der warmen Jahreszeit
darstellt. Diese Temperaturen sind nach
den Meridianen verschieden. So ist
z.B. auf der Isothermkurve von 10°,
im Küstenklima von West-Europa
die Temperatur des Winters 2°, des
Sommers 20°, im Kontinentalklima
von Ost-Asien, jene 5° diese 26°,
an der Ostküste der Neuen Welt; Temperatur des Winters - 1° des Sommers 22°.

TEMPERATUR

IN VERTIK. RICHTUNG

NEÜE WELT

Gezeichnet von Bär, Berghaus, u.Schelle.

Berghaus's Perspective

Heinrich Berghaus, one of the early scientifically motivated cartographers, published Humboldt's pioneering isothermal ideas as a world map in 1838. It is unclear if he was aware of Woodbridge's prior efforts. This 1849 edition by Berghaus, with German text, uses a polar projection to highlight the northern hemisphere's average temperatures, as well as the role of mountains and snow (listed and sketched by location and presumed height along the bottom).

Maps had obviously been around for a long time, even global maps. Mostly political or for navigation, they traced our understanding of geography and trade routes. But they were not generally applied to scientific questions, not used as tools of insight in and of themselves. The questions scientists were asking were only just then reaching the proper scale to take advantage of these new, global perspectives. With scientific mapping for perspective and the proper questions, global research could now take place. In these isothermal maps, for instance, comparisons between places are immediately identifiable—leading to insights in fields as disparate as deep evolution (Why do many plants look similar in Mexico and India?), biogeography (Why is tree line fairly consistent around the world at about 50°F [10°C] in the summer?), and disease (Why are there spatial patterns of malarial infections?). The isothermal temperature maps, predicated on a stable mean to create long-term patterns, were a gateway to global science.

We are now fully familiar with the global perspective of big maps, but today the concept of mean annual temperatures, presented in a way that implies consistency and stability around the world, seems almost quaint. Our ability to track temperature change—at high spatial and temporal resolution—means seeing the thermal world as a more dynamic place than simply as averages spread around the globe, as some places warm and others cool. Humboldt's isolines are getting tangled.

The atmosphere varies not just across space and in terms of averages (like the isolines), but at fine scales (valleys versus mountains) and in time. To capture the new thermal world, and the variability that we must also grasp to understand it, science now turns to spatially explicit computer modeling, which allows for complex extrapolations at high spatial and temporal resolution (for example, a prediction for every square kilometer for every hour of every day, something unthinkable in Humboldt's day). These methods are informed by a dense network of monitoring sites, satellites, and math. Such charting gives us dramatic pictures of a thermal world gone wild.

In January 2019, temperatures plunged in the eastern United States to North Pole levels. Climate change deniers used it as a reason to cry foul (yet again), saying the globe was not really warming. But that is a mistake of scale, a result of our limited daily perspective and willful ignorance of the global picture that we now possess. In fact, the cold weather was quite obviously simply a displacement; warm air normally in the south and cold air normally near the poles switched places as a consequence of the polar vortex spinning out of its usual place. The predictable

OUR ABILITY TO TRACK TEMPERATURE CHANGE . . . MEANS SEEING THE THERMAL WORLD AS A MORE DYNAMIC PLACE.

atmospheric circulation cells that set up the averages of Woodbridge and Humboldt are breaking down with the addition of heat into the atmosphere. At a local level, that can mean colder rather than warmer days—but the colder spots are generally offset by warmer spots elsewhere. During a similar polar air outbreak in 2018, Gulf Coast residents complained of unusually cold weather, but I lived in Alaska at the time—people were in T-shirts on the beach. Some make the mistake of thinking all weather is local; it is not. Weather is but a piece of the global scene, and only by taking in the whole can you get a sense of what is really going on.

Do small changes in temperature really matter?

One place you can really see the impact of rising temperatures on life and non-life is Glacier Bay, Alaska. Although several tidewater glaciers still calve (lose ice in large chunks) into saltwater, delighting cruise ship passengers and making frozen, ice-blue rafts for seal pups, the name "Glacier Bay" itself is rapidly becoming a misnomer as its

The polar vortex of early 2019 swept through eastern North America, bringing record cold to low latitudes. In this broadscale map showing the Arctic, one can easily see that this event was not a random blast of cold air but a rearrangement of cold and warm air; at the global scale, the winter was quite warm overall. Air temperatures at 6 ft. (1.8 m) in the Arctic Ocean warmed to near freezing as the normally frigid air slid south, incredibly warm for midwinter near the North Pole.

NORTH POLE

namesake glaciers recede and thin. The landscape is more trees than ice today. But it is not as simple as saying human-caused climate change is responsible for the disappearing ice. That is what makes Glacier Bay appealing as an example. Human-caused climate change now dominates, but there are a variety of global-scale natural factors and contexts that filtered the impact through the past. It is complicated, and that is what makes it interesting.

Glacier Bay was not originally a bay. In the oral history of the Tlingit, the prevailing Indigenous peoples that still call the region home today, it was a wide river valley and home to an excellent sockeye salmon run, so strong the river itself, the Ghathéeni, was named after the reds (sockeye are known locally as reds because of their intense red coloration when spawning—not to be confused with "redds," depressions fish make in the streambed where they lay their eggs). The main village was S'é Shuyee, the "area at the end of the glacial silt," likely near tidewater. The broad valley in which the village was nestled was headed by a glacier system born in the Fairweather Range of mountains, just to the north. In one telling, the men of the tribe decided to call the glacier down, which they did with a piece of king salmon, but the glacier came "like a tide." "Overnight," they were chased out of their homes, moving first to the mouth of the river, then to an island offshore (likely what is now called Pleasant Island). But it was still stormy, a fierce wind blowing from the ice sheet just a short distance to the north. Ice chunks broke off, causing waves to push up rocks, shells, and sea life into the woods around the village. Eventually, the narrative continues, the people moved farther south, to what is now Hoonah, Alaska (which means "a place of shelter from the north wind"). The story of the descending glacier (and other variants about the glacial advance) has been passed down for generations, and if you can make it to Hoonah, you should ask to hear it told by one of the original people.

> ONE PLACE YOU CAN REALLY SEE THE IMPACT OF RISING TEMPERATURES . . . IS IN GLACIER BAY, ALASKA.

The climatological cause of the ice sheet advance is not well understood, but current theories point to a simple temperature feedback with massive implications. One compelling hypothesis is based on volcanoes. Around 1250 CE, a series of four sulfur-rich volcanic eruptions in just 50 years spewed particulates and aerosols that swirled around our well-mixed atmosphere and cooled the globe rapidly by limiting solar heating. The effect was strongest in the northern hemisphere (on average, it appears the globe cooled only about 1.8°F [1°C]). Likely a result of the slightly cooler summers, winter ice did not melt off and ice caps expanded—leading to cooler temperatures as the bright white snowcaps bounced heat back into space.

August 10, 2018, was hot in the Pacific Northwest. This is a land surface (not air) temperature map of the day, highly relevant to fires, moisture stress, and plant survival. The extent of the heat is easily seen in the anomalies depicted here—areas of red are much hotter than average for the date; areas in blue, cooler.

Like some sort of frozen, self-fulfilling prophecy, each winter created a bit more ice, which made the world a bit colder, making even more ice the next year. All from less than 2°F of widespread cooling.

Other potential explanations exist and could fill a book themselves. They are not mutually exclusive. The Spörer and Maunder Minimums, both periods of low solar sunspot activity, could be partially responsible—though tying sunspot activity to global climate in a causal way is a difficult task. Low sunspot activity is correlated with slightly lower radiation from the sun, but tracing that force through the other major factors, like volcanos, is hard to do. Changes in ocean currents, specifically the movement of warm water to higher latitudes, has also been blamed. When fresh water moves into the ocean faster than normal, such as when glaciers melt, it influences how and where the ocean "turns over" as fresh water cools and sinks. A slowing of this movement would limit the amount of heat making it to high latitudes in the

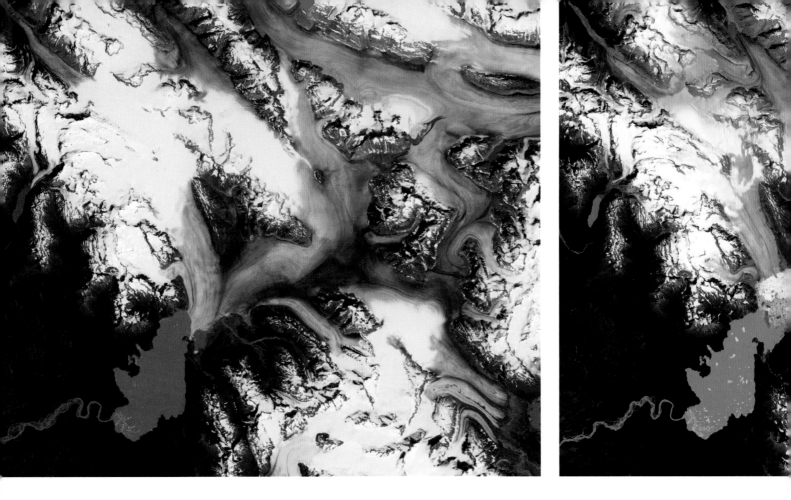

north, exactly where the glaciers were overrunning villages (more on that later). Finally, some propose that the Black Death (in the 14th century) and the depopulating of the Americas in the 16th and 17th centuries led to so much forest regrowth from abandoned agricultural land that global CO_2 levels dropped. In fact, there is a drop in CO_2 levels noted in Antarctic ice records from about that time. Regardless of cause, however, what is known as the "Little Ice Age" was a major disruptive event throughout the northern hemisphere.

George Vancouver sailed to the mouth of what is now Glacier Bay in 1794, not long after the high point of the advance that the Tlingit had dealt with earlier. A wall of ice, nearly 4000 ft. (1200 m) thick and around 12 mi. (20 km) wide spanned the bay. It was treacherous; the HMS *Discovery* was too large to maneuver through the ice-choked waters, so longboats were taken through the bergs (an area still named Icy Strait, despite no longer containing any ice). That was just the end of a river of ice 60 mi. (100 km) long.

Despite its magnificence, the ice was actually in full-fledged retreat at that point, about 6 mi. (10 km) back into the newly formed bay (the exact location of

The progressive loss of ice from the Yakutat Glacier near Yakutat, Alaska (left to right: 1987, 2013, and 2018) is evident from the declining white ice replaced by slate-gray water full of milky glacial runoff and land. To date, the retreat has spanned 3 mi. (5 km), and the glacier is thinner as well.

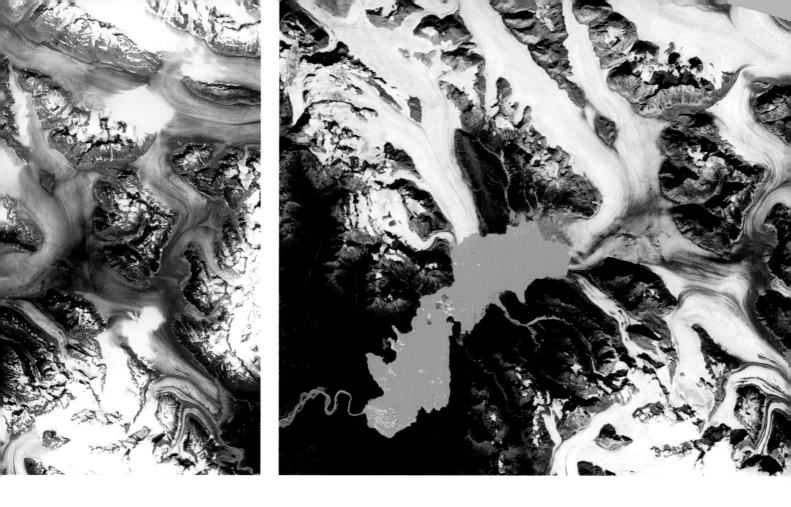

the edge at that time is still unknown). By 1879, when John Muir arrived and sang the praises of frozen hills, the ice sheet had split into two lobes, the western having retreated nearly 50 mi. (80 km) in those 83 years and the eastern, about 28 mi. (45 km). Nonetheless, he was stirred:

> As long as I live, I'll hear waterfalls and birds and winds sing. I'll interpret the rocks, learn the language of flood, storm, and the avalanche. I'll acquaint myself with the glaciers and wild gardens and get as near the heart of the world as I can.
> —John Muir, journal

Today, the ice has pulled back even farther; a lot of my work is in Glacier Bay and I have stood on the pile of rocks that was once Muir's cabin. He built it at the foot of the glacier that still bears his name—but now the ice has retreated so far it can't even be seen from his nearly forgotten stones.

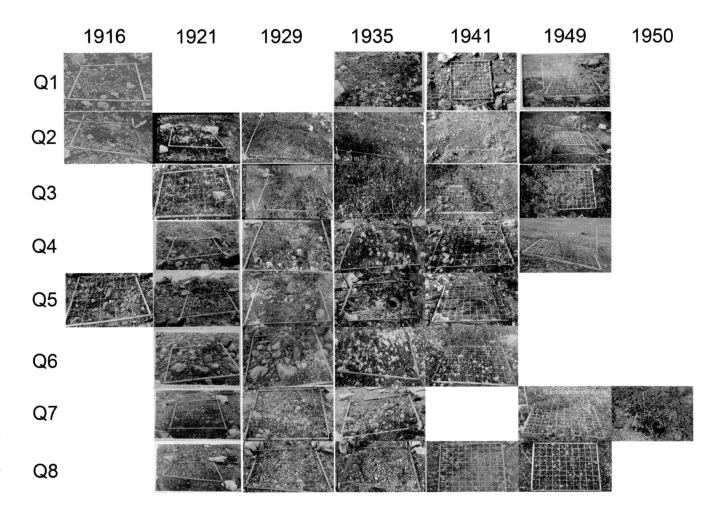

	1916	1921	1929	1935	1941	1949	1950

Q1 Q2 Q3 Q4 Q5 Q6 Q7 Q8

Was it global warming? Not in a simple way. The rapid retreat of the 1800s and early 1900s is primarily attributed to the tidewater glacial cycle. Tidewater glaciers are fascinating and illustrate the power of positive feedbacks to maintain processes once they are kicked off by some force—a small example of what's going on in our atmosphere all the time. As glaciers begin to advance into the sea, they push up a wall of sediment at the front, providing support to the front end and thus reducing calving substantially—while also providing resistance that allows the upper portions of the glacier to thicken. The glacier will continue to push itself, and that sediment, forward as upstream ice weighs down on the front, for decades or centuries. Eventually, however, the large mass of ice in lower, warmer conditions slows the advance as it gets too heavy to push. Once that happens, the glaciers stall for a time, as the lobe of sediment limits ice loss—but slow melt at the edges will eventually outpace the deposits. Once the melt causes even a slight retreat off that ridge of sediment, the tip of the glacier is back into the deep fjord water it has just excavated—and it will start

This image shows the growth of life in Glacier Bay over the last century. Starting from weeds and sprouts surrounded by glacial boulders, now the landscape is near-impenetrable forest and shrubs, frequented by bears and moose. These locations have been revisited every decade—via boat, then kayak, then foot—since 1916 and precisely remeasured, though the markers are now buried in new soil.

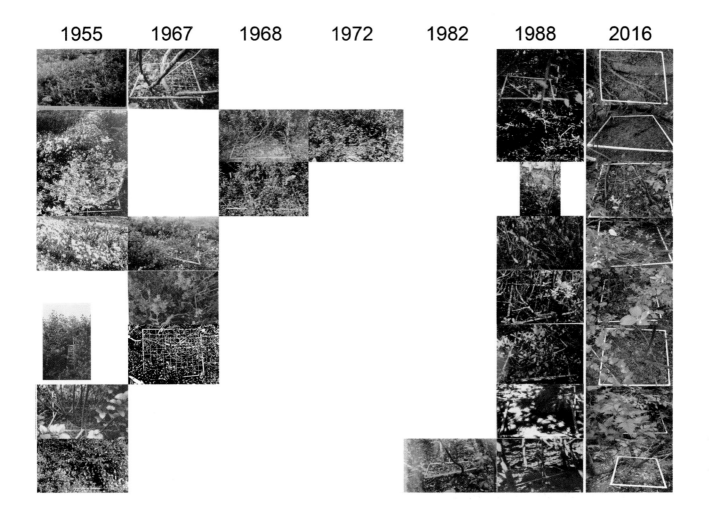

1955   1967   1968   1972   1982   1988   2016

calving off the now-destabilized end, with large bergs tumbling into the deep water as tides, weather, and temperature conspire to break the ice. Dramatic retreat begins as the now massively overextended ice river collapses into chunks of floating detritus—half a kilometer per year, maybe faster. Eventually the glacier retreats so far, it moves right out of the warm waters and elevations that drove the rapid shedding of ice, and the thing starts to grow again as upstream inputs outpace downstream losses. Accelerated growth can overextend in the same way as previously described, starting the cycle again (and this may also be part of the story of the Glacier Bay advance). This is a cyclical, physics-driven phenomena and Glacier Bay appears to have been caught in mid-cycle when Captain Vancouver sailed up. The retreat through the beginning of the 20th century was the last phase of this dramatic behavior.

But since the 1950s, the fingerprint of anthropogenic (human-initiated) warming has become unmistakable. Many glaciers of Glacier Bay have generally stopped retreating—that phase is done—but they have continued melting. They are becoming thin and spindly, now more rivulets of ice compared to the mighty rivers of John Muir's day. Rather than ice and rock, Glacier Bay is now dominated by plants. The shorelines are thick jungles of bushes, shrubs, and tree-lined wild strawberry beaches. This too is a fascinating story; the growth of the new ecosystem has also been charted for over 100 years. Careful monitoring of plots every decade or so has served as a clear witness to the transitions from weedy, post-glacial flowers to shrubs and forests. As the frozen system passes, the living system advances—and accelerates. With the warming climate, plants are growing faster than they have in past centuries. This increase in life as a result of climate change is an interesting counterpoint to losses seen elsewhere; it shows how the story of a place matters.

Charting the rise and fall of Glacier Bay can teach us a lot about what is, and what is not, a product of global temperature change—the gentle swings of natural cycles that can pass important thresholds and kick off dramatic fluctuations on the ground. Both human-caused climate change and natural systems can trigger change.

Blue Mouse Cove, under ice only 130 years ago, is now all forest and wild strawberries. It is also base camp for ongoing research into ecosystem development after glacial retreat (my tent can be seen in the shadows).

And all of those things are parts of the larger Earth system, and so they interact, changing the Hoonah Tlingit homeland several hundred years ago and the homelands of all us today.

PRECIPITATION: THE CHALLENGE OF COMMUNICATING VARIABILITY

The other major thing folks think about when they think atmosphere is precipitation, the companion of temperature in our weather forecasts. Knowing the *average* precipitation of a region is valuable, certainly, but what is generally more important is the *variability* in precipitation, both within and between years. Some areas get a decent amount of precipitation, but it is all within a few months, the wet season. Other areas get the same amount of precipitation but it is spread evenly throughout the year. And both the average amount and variability of that range is changing.

For example, some places are getting drier. But is a single dry summer a drought? The concept of a drought is closely related to anticipated variability because it refers to some sort of deficit. So we must tie the concept of drought to a particular place on the map, because the idea itself is relative to some expectation—a dry summer isn't necessarily a drought if dry summers are normal (but it could be if the preceding winter was abnormally dry). In other areas, where rainfall is typically plentiful all year round, a drought could be felt after only a month without measurable precipitation. As a result of this confusion, many scientists are moving away from using the word "drought," because it has so many connotations to different people. A modern term is "ecodrought," which forces us to think about the local ecosystem. Clearly the shrubs of the Arabian Peninsula will not feel a few months without water, but the thimbleberries of southeast Alaska would be crispy to the touch after that much time (and have been, recently). A common misconception is that climate change means drier conditions everywhere. That is not the case—in fact, net global precipitation should increase (estimated around 1.5–2 percent per 1.8°F [1°C] rise in surface temperatures), because warm air holds more water. Many places will get wetter, increasing flooding and its damages. Other locations will certainly get drier, as summer monsoons or winter storms shift their patterns elsewhere. A good number of areas will become more variable. But one certainty is that nearly *all* places are getting hotter—and as a result, the need for water is going up.

A COMMON MISCONCEPTION IS THAT CLIMATE CHANGE MEANS DRIER CONDITIONS EVERYWHERE.

Drought

Drought is difficult to visualize. It is both a spatial and temporal phenomenon, and expressing two types of variability is challenging. This map displays 285 weeks of drought data in the US. Dot size represents the amount of time in those nearly 5.5 years that was spent in drought conditions (the largest being 100 percent of the time) and color displays the intensity of the drought, with purple being the most intense. Chronically dry areas have large dots, and areas that were exceptionally dry during that time have both larger dots and darker colors, highlighting the spatial variation even within droughty conditions.

Areas of the Shoshone mountains have spent 82% of the past five years in drought; over half of that at "exceptional" intensity.

Parts of the Pinaleno Mountains, in AZ, have sustained drought over the *entirety* of the past five years.

Frequent exceptional drought

Infrequent, though exceptional, drought

Frequent, though moderate, drought

Severity
Weighted proportion of drought intensity

Frequency
Proportion of time, over five years, spent in drought

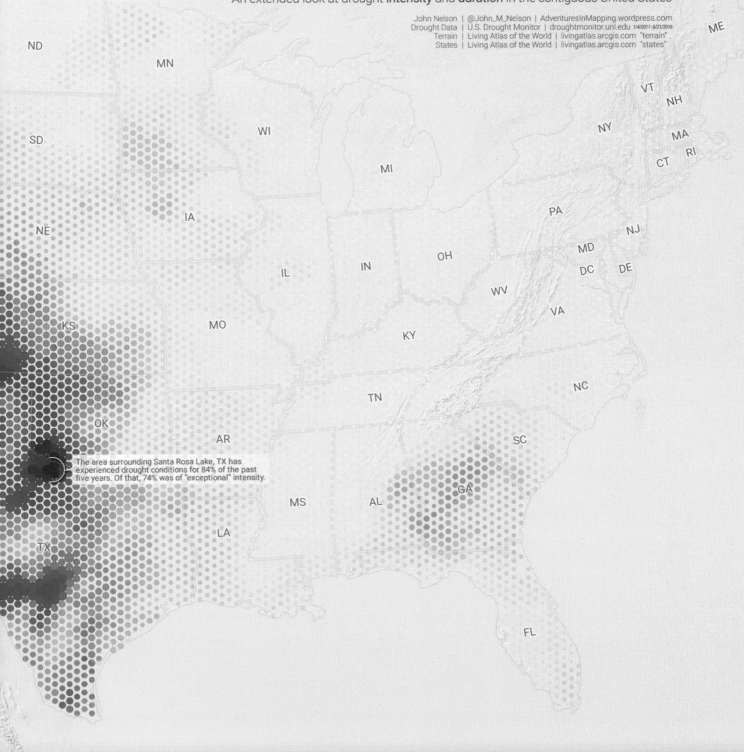

Five+Years of Drought

An extended look at drought **intensity** and **duration** in the contiguous United States

John Nelson | @John_M_Nelson | AdventuresInMapping.wordpress.com
Drought Data | U.S. Drought Monitor | droughtmonitor.unl.edu 1/4/2011-6/21/2016
Terrain | Living Atlas of the World | livingatlas.arcgis.com "terrain"
States | Living Atlas of the World | livingatlas.arcgis.com "states"

The area surrounding Santa Rosa Lake, TX has experienced drought conditions for 84% of the past five years. Of that, 74% was of "exceptional" intensity.

In very broad strokes, temperate latitudes (say 35–50 degrees north and south) will probably dry; lower and higher latitudes will likely get wetter. Everywhere will warm—the highest latitudes fastest, the lower latitudes more slowly. Temperatures will also swing more wildly as historical patterns in atmospheric circulation are weakened or upended and as feedbacks—such as the icy poles reflecting heat back into space—are lost. The effects of warming and precipitation change will intertwine in countless ways, making many places more pleasant and many places much less so. But what is ultimately true is that treating them separately is misleading because one thoroughly influences the other. It is impossible to do a complete treatment of all those mixed effects in a book, much less a chapter, so here we will pick a case study that will be familiar to many.

Disappearing snow

Precipitation and temperature combine to generate one of the most beautiful of natural phenomena on the planet: snow. However, as many a school child can attest, there is a painful difference between 32 and 33°F (or 0 and 1°C). That one degree is big. It is the difference between freezing and thawing, snow days and school days. While most of the talk about climate change involves several degrees over 100 years, one of the biggest impacts is hidden within that single degree of warming.

Disappearing snow can be traced in a variety of ways, from trends in averages to increases in variability (and resultant years with little to no coverage). Here, the Global Historical Climatology Network, which provides weather records stretching back to 1763 in some locations, shows those changes clearly. A word of caution: the dataset is built from verified ground observations, which are primarily found in places with official weather stations, mainly cities, major highways and infrastructure installations, and airports. This bias can result in some differences when compared to other datasets, such as those derived from remote sensing (which has its own challenges, including clouds). Data shown here represents an average of stations within approximately 93 mi. (150 km) and days with more than 1 in. (2.5 cm) of snow.

Around the Great Lakes, the number of days of snow cover has declined.

At weather stations throughout the Rocky Mountains, the number of days with snow on the ground is high and varies relatively little year to year.

Throughout the Northeast & Great Lakes, there can be a big difference in snow cover from year to year.

YEAR-TO-YEAR SNOW COVER VARIABILITY

Outer ring size: Most days of snow cover in a single year between 1995 and 2019

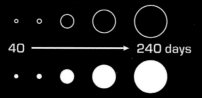

40 ——————→ 240 days

Dot size: Fewest days of snow cover in a single year between 1995 and 2019

CHANGE SNOW COVER DAYS

Color: Change in the typical number of days with snow cover from the present (1995–2019 average) to the past (1958–1982 average)

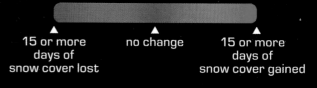

▲   ▲   ▲
15 or more   no change   15 or more
days of   days of
snow cover lost   snow cover gained

Hatching: Areas with sparser weather data

Like Alice, take a drink of some bottle best left alone and shrink down till grass towers above your head. Rocks are mountains, lawns are forests. Everything is great in the summer—seeds are gigantic feasts and puddles are lakes. You are very small, so you need less food and the ground is a warm and inviting bed at night. Summer is a time of plenty. But pretty soon, the days shorten, the shadows lengthen, and it gets colder. Much colder. (This is not only an average winter temperature thing—your mouse-sized self gets cold much faster than your larger self. Smaller animals have a higher surface area relative to their body, which means they radiate heat faster from their core. You can see this in action; although there are exceptions, many animals in colder environments tend to be larger than their warmer-climate counterparts, a phenomena known as Bergmann's rule. Bigger animals have a lower ratio of surface area to volume, meaning more heat is retained in the body.) Some animals hibernate or enter a state of torpor for the cold weather, but many do not. Rats and mice tromp around happily even when the thermometer plunges below zero—the tracks in the snow surrounding my chicken coop testify to that! This is where snow comes in. It turns out snow is an incredibly effective blanket, and while the atmospheric temperatures may drop far, far below freezing, temperatures at ground level when there is a layer of snow are often quite warm (relatively), often hovering right around freezing.

The technical term for the under-the-snow-but-above-the-ground environment is the "subnivium," and it is a valuable habitat in and of itself, providing another way in which species can specialize and biodiversity can be maintained.

Plants take advantage of this blanket as well—rooting zones, especially in cold climates, are kept warmer in the winter than local air temperatures would suggest by this insulation from the freezing atmosphere. Ample life avoids freezing to death simply by staying low, close to the ground, and under the snow. Warming strips away that insulating white blanket, and thus has a paradoxical effect on creatures reliant on the subnivium or plants that depend on insulated soil. Global warming is making their world colder.

Yellow cedar, the canary of the woods

On the remote North Pacific coast of North America you can really get away from people. I lived there for many years and continue to do most of my research on the misty fjords and bear-track-strewn beaches of southeast Alaska. You often see more bears than people and more eagles than bears. Salmon runs are still strong after 10,000 years. The weather arrives slowly and steadily from the west after several thousand miles of open ocean. It is almost always raining, and the drops feel deeply, intrinsically clean as they drip off the spruce and cedar branches. My old, mossy house in Juneau, Alaska, received more than 78 in. (2 m) of precipitation a year, evenly spread throughout the winter and summer. It has done this for several thousand years, at least; every winter, every summer. It's incredibly, deeply damp all the time. The plants are all well adapted to the wet; the ground nearly saturated throughout the year. But the precipitation is nearly all rain now, and that is new. And the forests are dying as a result.

Although southern Alaska is as far north as parts of Norway and Siberia, most of the winter is spent at or above freezing, a result of the warm ocean currents that meander up from California, Oregon, Washington, and British Columbia. And recently, it has been getting warmer. Not a lot—the ocean itself does a good job buffering both high and low temperatures. It is not sheer heat nor any change in the amount of precipitation that has killed nearly one million acres (400,000 ha) of yellow cedar trees across 10 degrees of latitude. But something turned whole islands from feathery yellow-green to standing gray sticks.

Questions of pathogens, like fungal infections, came up first, then insects. But tests were negative and the root cause unknown. The break in this cold case finally came from maps of snow. The dead trees were concentrated at lower elevations and on the outer coast. The real kicker came when researchers overlaid a map of dead cedars on a map of areas with extensive snow loss—regions where warming had changed the winter precipitation regime from snow to rain, even though the overall precipitation amount had not changed. The areas were a near-perfect match. Although winter temperatures have only warmed a few degrees since the mid-1800s in most places along the coasts of British Columbia and southeast Alaska, they have crossed that critical threshold of 32°F (0°C). If you map those places, you have mapped the tree killer.

THAT SEEMS A STRANGE THING. WHY WOULD LESS SNOW KILL TREES?

That seems a strange thing. Why would less snow kill trees? Yellow cedar is one of those species that specializes in snow and that subnivium. It "wakes up" from winter earlier than most trees. This is an adaptive strategy; all the early-season resources are available for the taking. But it also exposes a vulnerability: fine roots, metabolically active, are highly susceptible to freezing damage. And while the world is warmer, cold snaps still occur, partially attributable to the instability in atmospheric circulation mentioned earlier. Those cold snaps can freeze the soil that is bereft of its insulating white blanket, killing the shallow roots of the yellow cedar. Left without the ability to take up water, the 130-ft. (40-m) trees struggle and die, sometimes immediately, sometimes over a period of years. A short cold snap takes out giants.

We do not know if other species around the world are similar; southeast Alaska is one of the first to lose essentially all of its winter snowpack on a broad scale (though

Receding Cedar

The loss of snow on the North Pacific coast of North America is made manifest by the swath of dead trees left behind, as roots freeze in the newly exposed spring soils. From sea level in the north to higher elevations in the south, the line of lost snow is slowly sliding north, claiming forest as it moves with the warming climate.

This graphic shows the progression of disappearing snow and its consequence: disappearing trees. Snow loss along the North Pacific coast of North America is causing mass mortality among yellow cedar, a culturally significant tree. Dead zones (in red) follow the line where winter temperatures are shifting from below freezing to above freezing, from snow to rain.

Colder locations at high elevations and warmer locations where there are no winter freezes are generally healthy (in green).

JUNEAU, ALASKA

N

**VANCOUVER,
BRITISH COLUMBIA**

The southernmost areas, where
the species is at it highest numbers,
have no snow-loss-related mortality.

**SEATTLE,
WASHINGTON**

PORTLAND, OREGON

Dead trees are found
at higher elevations in the
mountains moving south.

Mortality is at
sea level in the north.

Scale varies in this perspective. Data from Buma 2018.

there are still snowy winters, of course—variability goes both ways even when there is a trend toward complete snow loss). There could be other vulnerable species in areas where snow packs are thinner and thinner each year, unaware that their time is coming as winter temperatures approach that critical threshold and their world will fundamentally change.

CHANGES IN THE WINDS

The airy parts of the world are a study in how the scale and scope of the variability and change in the natural world defies the scale of human experience. The issue is that the atmosphere is so incredibly big and all-pervasive. Small changes in average conditions have incredible impacts through sheer scale. A small uptick in concentrations of a rare gas, measured in parts per million and only 0.04 percent of the air itself, sends the world's climate into chaos. Volcanic eruptions slightly increase the concentration of sulfate molecules and cool the planet, encasing parts of the northern hemisphere in ice and pushing people from their homelands. A few droughty years make the difference between a successful homestead and a climate refugee; a few warm winters can kill off a forest. The natural variation of the atmosphere is large, but life's tolerance for those swings is finely tuned. Perturbing long-established conditions and patterns is a dangerous undertaking.

The steady upward trend in temperatures is shifting the average into ever-warmer territory. That drags the variance along with it, sometimes even destabilizing long-standing relationships between dynamic processes that limit the variation. Many are surprised that recent record-cold winters in their hometowns can be tied to global warming, but it makes perfect sense when one thinks about the ebb and flow of the atmosphere and considers the global perspective. The interior of a continent is intimately related to the conditions on the coast; there is no such thing as separation or isolation when we talk about the atmosphere. The clouds floating over your head were shaped by processes thousands of miles away—and locally. The systems are just so big—colder conditions in one place joined at the hip to warmer temperatures elsewhere—that the global view, truly a requisite one when considering these phenomena, is the only way to approach the atmospheric world. To imagine that we can manipulate this system but somehow be isolated from the effects is pure foolishness. But that is what we are doing, a grand experiment with an unknown outcome.

The complexity of the atmosphere, the feedbacks and chemistry, the scale, and the dynamic interplay between land, water, and air makes for a very hard-to-predict system. We know many of the potential changes are serious. Certainly changing average conditions are a concern, slightly drier years mean chronic water stress for thirsty cities. Equally or more concerning is that variability—the potential for wild swings to stress our ecological or human systems past the breaking point. Once broken, things don't always recover. Even average conditions that can support some type of life, say a forest ecosystem, may not be enough if the variability is such that in one year those conditions don't support such life; death is a point that a tree reaches only once, even if things get better. Finally, there are thresholds of sudden change we may not even be aware of. The shift from snow to rain in the winter is just such a point, with only a slight warming of temperatures causing a fundamental change in conditions. Few suspected that the loss of snow would lead to widespread forest mortality due to roots freezing, but that is exactly what happened. These hidden thresholds are hard to even imagine, much less anticipate.

THE CLOUDS FLOATING OVER YOUR HEAD WERE SHAPED BY PROCESSES THOUSANDS OF MILES AWAY—AND LOCALLY.

The seemingly ephemeral nature of the weather, plus our limited scale of observation on the ground, can lead many to assume the atmosphere is always changing, so variable that small shifts in average temperatures shouldn't be a big deal. If the weather forecast is off by three degrees and it isn't the end of the world, why should we worry about a change of 1 or 2 degrees over 30 years? That perspective is doomed because of scale, the sheer size of the atmosphere as a single system.

It is like the proverbial blind men studying an elephant, each feeling a different part and arguing about the whole; the local info may be interesting but you will never get the full picture without expanding your view. Charting and mapping the atmosphere is beginning to give us the full picture, though, or at least enough to open up our perspective to one that considers the system as the single, dynamic pool that it really is. With this wider, better viewpoint comes the realization that we simply cannot ignore other, distant parts of the world, because when it comes to the atmosphere, there are no truly *other* parts of the world.

WATER

THE GLOBAL BUFFER

Water is the new oil—a common sentiment found in punditry, economic predictions, and conservation circles. It comes from the simple reality that water is perhaps the most precious natural resource we have. Simply put, unlike oil or gold, water is necessary to sustain life itself. All living organisms need access to water, and unless you live by the ocean or a lake, that water can be variable, intermittent, or hard to come by. Water-blessed regions enjoy their liquid security; water-poor areas send out tendrils in the form of pipelines, canals, wells, and waterways to siphon off enough to survive dry years. And we have mined it like gold, as we'll discuss in the next chapter. Conflicts over water, and the lack of it, are likely to be one of the main precipitators of conflict throughout the 21st century. Take the oceans: while not typically used for drinking, they are the great protein factories of the world, long supporting fisheries that have sustained billions. Although fish and protein resources were diminished by mechanized overextraction around 75 years ago, and curbs were put on yields (with varying degrees of effectiveness), fisheries remain major sources of protein today. Trade is conducted primarily by sea. There is no escaping the reality that water is our most precious resource.

"We are tied to the ocean. And when we go back to the sea, whether it is to sail or to watch—we are going back from whence we came," said John F. Kennedy. But even these benefits are dwarfed by the climate regulation provided by the world's waters.

OCEANS

The oceans play an immense role in climate management, buffering the atmosphere from our carbon emissions. Push them too far, though, and the oceans push back. Despite the cushion they give us against climate change, the steady and interminable warming of the atmosphere is adding water to our oceans from the ice caps of the world, slowly now, but accelerating. The ocean is itself warmer, expanding inexorably like metal on a warm day. The chemistry of the air is seeping into the waters. And as all that happens, the ocean is revealed as an active player in the story of climate change—slow to get moving, but one that is starting to inexorably claim land from under our feet and houses and to change in fundamental ways.

The constant, unchanging nature of the oceans and waters seems the antithesis of the mercurial atmosphere. Tempestuous, yes. A summer squall or winter blow is to be expected. But overall, the ocean seems steady, perpetually itself. Yet that too is an illusion brought on by the mismatch in scale between our limited perspective and the system that is Earth's waters. Just as the weather's local chaos hides the deep atmospheric connections we all share, so, too, the slow rate of change in the ocean and the world's hydrology hides the dynamic nature of that system, one that rises and falls, warms and cools, and circulates. The illusion of stability hides the quiet but powerful momentum that is building beneath the waves. Like the air, it is all connected. "The least movement is of importance to all nature. The entire ocean is affected by a pebble," wrote Blaise Pascal. And we are throwing a lot of pebbles.

The oceans cover 71 percent of the world, and Earth is fortunate it is a watery planet. Dark blue water absorbs solar energy rapidly (about 92 percent of the sun's energy is absorbed, depending on water clarity), then flows poleward, bringing the

RIGHT   The ponderous movements of surface ocean currents and gyres— slow by the standards of the atmosphere—are massive. Driven by the heat of the sun, Earth's rotation, and differences in water density, they also plunge down into the depths, forming deep ocean currents that complete a long, slow water cycle around the globe. This water movement has buffered us from much of the heat of global warming as the warmed waters disappear into the vast abyssal plain, but those areas are now noticeably warming as well.

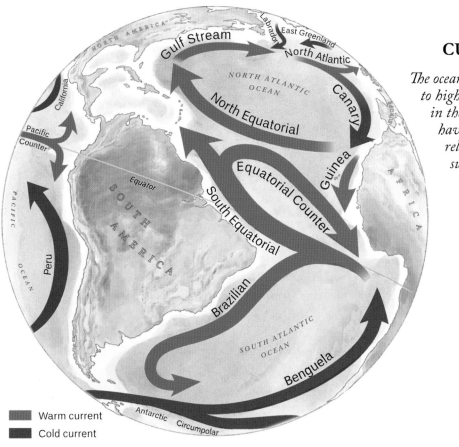

RIVERS IN THE OCEAN: CURRENTS OF THE WORLD

The ocean flows in mighty rivers, circulating heat to high latitudes and cool waters to the equator in the great gyres of the world. These currents have been used for centuries by sailors to reliably cross the vast seas, little boats surrounded by oceanic rivers.

Warm current
Cold current

SEA SURFACE TEMPERATURE

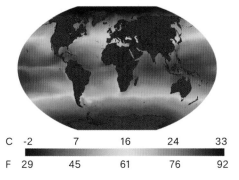

C	-2	7	16	24	33
F	29	45	61	76	92

FOLLOWING LEFT Navigating the North Atlantic required atmospheric as well as oceanic charting. This early data visualization from 1853 shows the direction of winds for every 5 × 5 latitude-longitude square of the North Atlantic Ocean, by month. Numbers within each circle show the days with winds in that direction. To make this map, Matthew Maury consulted log books as old as 1310 CE.

tropical warmth with it. North and south of about 35 degrees latitude (and depending on the season), the planet is losing heat to space, running a negative energy balance. Less energy in than out. The reason the temperature does not drop below zero, like any account spending more than it brings in, is heat transport—partially by the atmosphere and partially by the ocean. Some heat that enters the ocean is also circulated down to the depths, one of the reasons the ocean is such a strong buffer against climate change. But a significant amount of heat is transferred to the atmosphere via evaporation and radiation, especially when winds are strong or the air is dry. This heats the surrounding land, especially areas downwind of warm-water currents like the Gulf Stream. Meanwhile, the tropics continue to bring in a surplus of heat. As a result of this surplus and transport, the world—rather than just a strip around the equator—is warm.

Earth's water mass also provides a huge protective capacity against climate change. In the next section, we'll run through some basic consequences of climate change on the oceans' physical system, primarily a function of temperature and chemistry. There are many more "downstream" impacts on other facets of the ocean: the

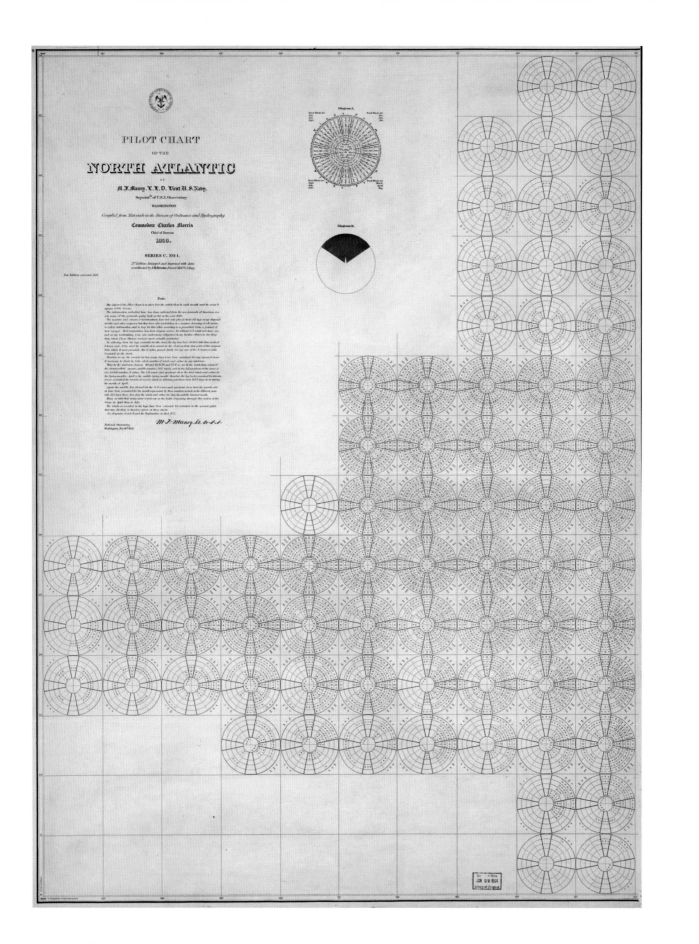

PILOT CHART

OF THE

NORTH ATLANTIC

BY

M. F. Maury, L.L.D. Lieut. U.S. Navy.

Superint.ᵗ of U.S.N. Observatory

WASHINGTON

Compiled from Materials in the Bureau of Ordnance and Hydrography

Commodore Charles Morris

Chief of Bureau

1852.

SERIES C. NO 1.

2ᵈ Edition Enlarged and Improved with data

re-collected by J.M.Brooke, Passed Mid.U.S.Navy.

New Edition corrected 1857.

The Grand Banks

The oceans have long been a bounty of food and resources. The Grand Banks off eastern Canada were once a massive supplier of the world's protein, and maps like this one, from the early 1800s, drew sailors from around the world to harvest cod in the region (note the "Nova Scotia Fishing Banks" labeled in the lower center of the map). After 500 years of successful fishing, the introduction of mechanized, large-scale fisheries in the mid-20th century decimated those populations, however, reducing cod numbers to less than one percent of historical norms. Fishing has been prohibited or restricted for decades, and populations are only slowly starting to recover.

migration patterns of whales, the life of coral, the survival of estuaries and the people that depend on them. But at its heart, climate change and the oceans are physical systems that function via fairly simple, physical processes: energy and acid.

Sea level and ocean heat

Osaka, Japan. Nestled by a large bay in the south-central part of the country, Osaka hosts the popular Tenjin Festival, an annual event with fireworks and a river parade. Here, it is projected that a rising sea will claim the land of 5.2 million people.

Alexandria, Egypt. Located at the mouth of the Nile River, Alexandria boasts a population that tops 5 million and once harbored one of the wonders of the ancient world, the (now-lost) Lighthouse of Alexandria. Here, the prediction is that an area now holding 3 million of the city's people will be below sea level.

Shanghai, China. The trade nexus of the world, Shanghai is the world's busiest container port and home of the Shanghai Stock Exchange. Though struggling with seawalls against the rising tide, as many as 17.5 million people are forecast to be flooded.

At a global scale, the logic is simple: sea level is rising and as a result, lowland areas are slated to flood over the next decades to centuries. That the world's largest cities are generally at sea level, on low-lying estuaries, is a natural consequence of how humans build their civilizations—the ecology of cities (more on that to come). Nearly 10 percent of the world's population (about 680 million) live in these low-lying coastal areas. But to paraphrase Tolstoy, each unhappy, flooded city is unhappy in its own way. The unique geographies and histories of each city make for equally unique situations that shape the impact and response, from the success of seawalls in blocking tides to the economics of the adaptations.

Miami, Florida. The glittering spectacle on the southeastern tip of Florida, the 8th-largest metropolitan area in the United States and the major economic powerhouse of southeastern North America. Florida is the poster child for the impacts of ocean change, and with good reason. The *highest* point in central Florida is less than 350 ft. (106 m) above sea level. Miami-Dade County is, on average, only 6 ft. (1.8 m) above the waves, about the height of the average adult; its highest point is only about 40 ft. (13 m) above sea level. Let's investigate Miami's story, and, like Glacier Bay, delve into the specifics that make it unique, and the generalities that make it instructive.

The dynamic story of the ocean and Florida starts over 300 million years ago. Florida enters this story as part of Africa. The deep basement of Florida is still

CLIMATE CHANGE
AND THE OCEANS
ARE PHYSICAL
SYSTEMS THAT
FUNCTION VIA . . .
ENERGY AND ACID.

REMARKS

Upon the Navigation from

NEWFOUNDLAND to NEW-YORK,

In order to avoid the

GULPH STREAM

On one hand, and on the other the SHOALS that lie to the Southward of Nantucket and of St. George's Banks.

AFTER you have passed the Banks of Newfoundland in about the 44th degree of latitude, you will meet with nothing, till you draw near the Isle of Sables, which we commonly pass in latitude 43. Southward of this isle, the current is found to extend itself as far North as 41° 20' or 30', then it turns towards the E. S. E. or S. E. ¼ E.

Having passed the Isle of Sables, shape your course for the St. George's Banks, so as to pass them in about latitude 40°, because the current southward of those banks reaches as far North as 39°. The shoals of those banks lie in 41° 35'.

After having passed St. George's Banks, you must, to clear Nantucket, form your course so as to pass between the latitudes 38° 30' and 40° 45'.

The most southern part of the shoals of Nantucket lie in about 40° 45'. The northern part of the current directly to the south of Nantucket is felt in about latitude 38° 30'.

By observing these directions and keeping between the stream and the shoals, the passage from the Banks of Newfoundland to New-York, Delaware, or Virginia, may be considerably shortened; for so you will have the advantage of the eddy current, which moves contrary to the Gulph Stream. Whereas if to avoid the shoals you keep too far to the southward, and get into that stream, you will be retarded by it at the rate of 60 or 70 miles a day.

The Nantucket whale-men being extremely well acquainted with the Gulph Stream, its course, strength and extent, by their constant practice of whaling on the edges of it, from their island quite down to the Bahamas, this draft of that stream was obtained from one of them, Capt. Folger, and caused to be engraved on the old chart in London, for the benefit of navigators, by

B. FRANKLIN.

Note, The Nantucket captains who are acquainted with this stream, make their voyages from England to Boston in as short a time generally as others take in going from Boston to England, viz. from 20 to 30 days.

A stranger may know when he is in the Gulph Stream, by the warmth of the water, which is much greater than that of the water on each side of it. If then he is bound to the westward, he should cross the stream to get out of it as soon as possible.

B. F.

The Gulf Stream

The Gulf Stream transports an immense amount of heat from subtropical latitudes to the North, warming continents and driving global climate. This well-known current, originally used in navigation for both trade and fisheries, has been a significant charting challenge for centuries, as this early map indicates (circa 1786 by James Poupard, with notes by Benjamin Franklin). A sailor wishing to go south would know he was in the Gulf Stream "by the warmth of the water, which is much greater than that of the water on each side of it. If then he is bound to the westward [not sailing toward England], he should cross the stream to get out of it as soon as possible," wrote Franklin. (Note that the Grand Banks of Canada are highlighted on this chart as well, above the Gulf Stream on the right.)

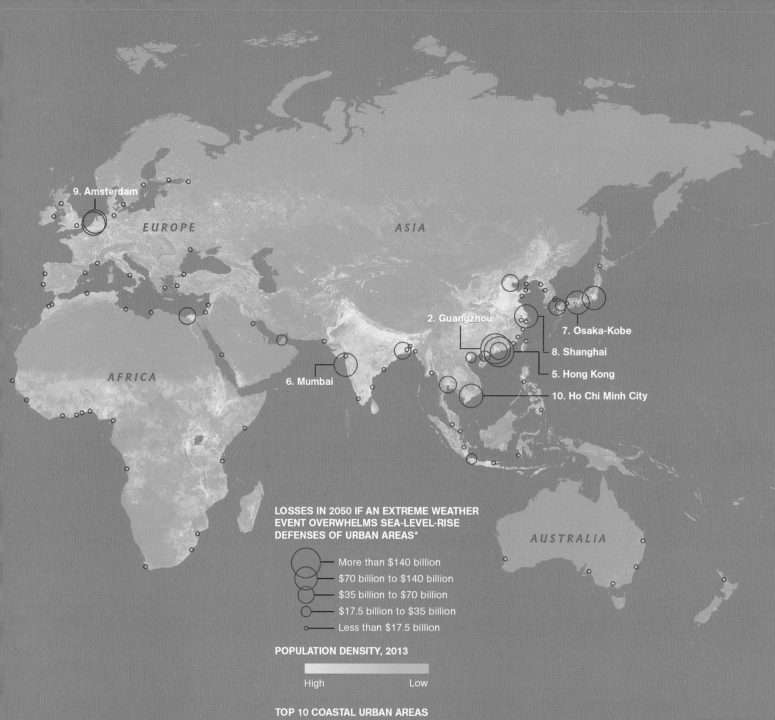

9. Amsterdam

EUROPE

ASIA

AFRICA

2. Guangzhou

7. Osaka-Kobe

8. Shanghai

5. Hong Kong

6. Mumbai

10. Ho Chi Minh City

AUSTRALIA

LOSSES IN 2050 IF AN EXTREME WEATHER
EVENT OVERWHELMS SEA-LEVEL-RISE
DEFENSES OF URBAN AREAS*

More than $140 billion

$70 billion to $140 billion

$35 billion to $70 billion

$17.5 billion to $35 billion

Less than $17.5 billion

POPULATION DENSITY, 2013

High   Low

TOP 10 COASTAL URBAN AREAS

1	Miami	$278 billion
2	Guangzhou	268
3	New York-Newark	209
4	New Orleans	191
5	Hong Kong	140
6	Mumbai (Bombay)	132
7	Osaka-Kobe	108
8	Shanghai	100
9	Amsterdam	96
10	Ho Chi Minh City	95

*ASSUMES CITIES CONTINUE TO BUILD PROTECTIONS ON PACE WITH SEA-LEVEL RISE TO MAINTAIN A
CONSTANT RELATIVE RISK OF FLOODING (IN 2005 U.S. DOLLARS) RYAN MORRIS, AND ALEXANDER STEGMAIER,
NGM STAFF. SOURCE: STÉPHANE HALLEGATTE, ET AL., *NATURE CLIMATE CHANGE*, SEPTEMBER 2013

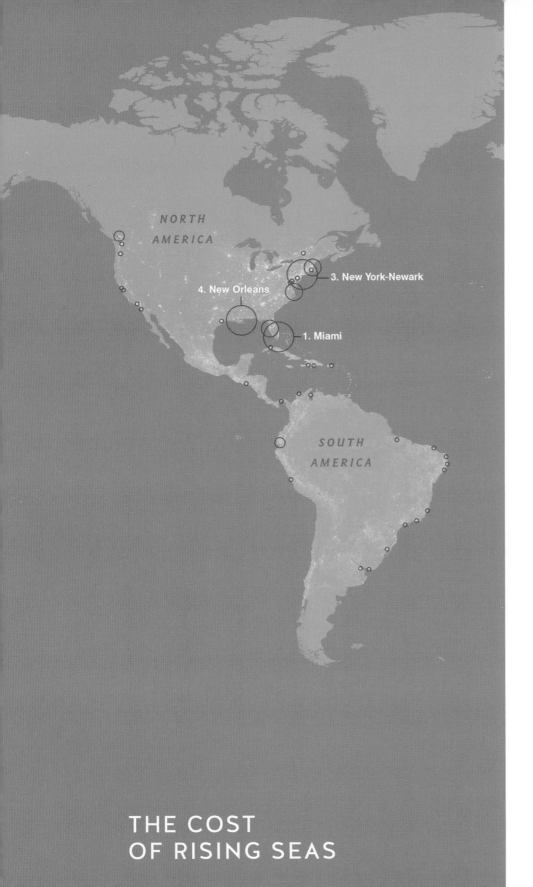

NORTH
AMERICA

4. New Orleans

3. New York-Newark

1. Miami

SOUTH
AMERICA

THE COST
OF RISING SEAS

Estimates of financial
losses due to sea level
rise in urban, coastal areas
by 2050 (in US dollars).
As coastal cities continue
to grow, especially in
South America and Africa,
more and more will be
put at risk.

African—more precisely, Senegalese. When the wandering tectonic plates came together in the mad crash that generated the ancient supercontinent Pangea, their edges fused in what is called a suture zone (it is still there, below southern Georgia). Like many a young couple, though, Pangea broke up. Planetary heat welling up into the center of the supercontinent eventually generated rift zones—far from Florida, not following the previous suture lines—that split the landscape down the middle and left Florida attached to its new dancing partner, North America.

So Florida is not just the subtropical terrain we see today, but rather a large chunk of Africa with some rubble on top. (In fact, the Florida peninsula, the south-ernmost tip of which Miami currently marks, is already two-thirds underwater. East to west, the "Florida platform" is about 155 mi. [250 km] wide, dropping down nearly 10,000 ft. [3000 m] on either edge, in dramatic escarpments.) The African rock newly ensconced on the edge of North America found itself almost entirely flooded due to warmer climates and higher sea level. But not deeply flooded, just enough for a shallow marine ecosystem to thrive, a sub-tropical environment perfectly suited for life. For millions of years, this highly productive seascape was home to billions of marine organisms, whose shells piled on top of each other, creating a thick carbonate layer on top of that African backbone. Florida was built underwater.

THE PENINSULA IS COMPLETELY AT THE MERCY OF SEA LEVEL.

Then, about 34 million years ago, Florida emerged (barely) from the waves. As the carbonate rocks surfaced, they began to dissolve thanks to rainwater, which is slightly acidic due to carbonic acid from the CO_2 in the air. The flat topography doesn't help, as water pools and ponds rather than slipping easily off. Like marble statues exposed to the more intense acidic rain that results from air pollution, the foundation of future-Florida began to develop pits and channels, creating what are known as karst formations. These formations in turn led to caves and springs, underground rivers, and the vast Florida aquifer which is now the source of drinking water for millions. Finally, and only recently (as in the last few tens of millions of years), erosional material from the southern Appalachian Mountains settled over the karst and carbonate, creating the sandy peninsula that we see today. This sand holds a lot of water as well, and so the vast peninsula, barely above the saltwater (remember, Miami-Dade County averages only 6 ft. or 1.8 m above sea level), is able to supply a population of nearly 22 million people with both land and fresh water.

What does this all mean for Florida? We have a long, nearly level peninsular ledge sitting on a broad, flat platform, built under shallow seas. The peninsula is completely at the mercy of sea level; the formerly high waters are what built it. And the

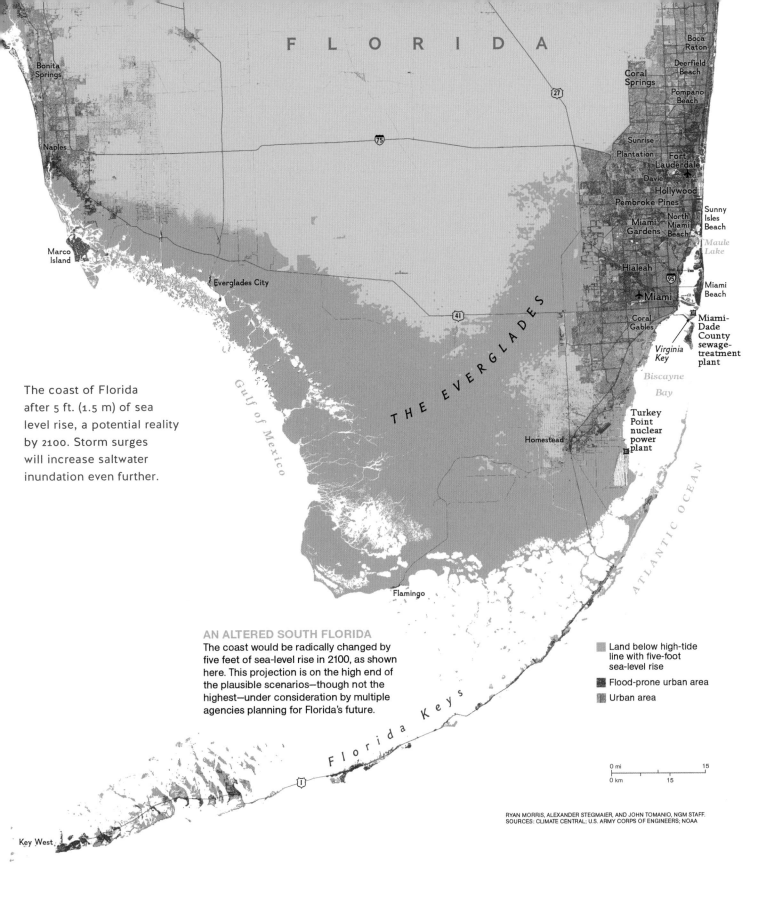

FLORIDA

Boca Raton

Bonita Springs

Deerfield Beach

Coral Springs

Pompano Beach

27

Naples

75

Sunrise Plantation

Fort Lauderdale

Davie

Hollywood

Pembroke Pines

Sunny Isles Beach

Marco Island

Miami Gardens

North Miami Beach

Maule Lake

Everglades City

Hialeah

95

Miami Beach

Miami

Miami-Dade County sewage-treatment plant

Coral Gables

Virginia Key

Biscayne Bay

The coast of Florida after 5 ft. (1.5 m) of sea level rise, a potential reality by 2100. Storm surges will increase saltwater inundation even further.

T H E   E V E R G L A D E S

Turkey Point nuclear power plant

Gulf of Mexico

Homestead

A T L A N T I C   O C E A N

Flamingo

AN ALTERED SOUTH FLORIDA
The coast would be radically changed by five feet of sea-level rise in 2100, as shown here. This projection is on the high end of the plausible scenarios—though not the highest—under consideration by multiple agencies planning for Florida's future.

Land below high-tide line with five-foot sea-level rise

Flood-prone urban area

Urban area

Florida Keys

1

0 mi   15

0 km   15

RYAN MORRIS, ALEXANDER STEGMAIER, AND JOHN TOMANIO, NGM STAFF.
SOURCES: CLIMATE CENTRAL; U.S. ARMY CORPS OF ENGINEERS; NOAA

Key West

ocean can pollute those freshwater reserves with salt percolating into the limestone, or take away the land entirely. In the next 15 years alone, property and development worth $15 billion is in danger of flooding.

Let's get one thing straight. As the climate warms, the oceans will rise. It is not complicated. There are two things going on: more water is being added to the ocean and the water is heating up, causing it to expand. It is pure and unforgiving physics that will likely swamp Miami, large parts of south Florida, and vast amounts of shoreline around the planet.

Much of the current and future rise in sea level will be caused by melting ice on land simply adding more water to the ocean. Water solidly above the waves as terrestrial ice is now draining into the seas. And while some high, icebound locations are seeing more snow as warmer air brings more precipitation, most sites aren't. The largest sources of this new water are Greenland, Antarctica, and mountain glaciers in Alaska and the Himalaya—about 0.07 in. (1.81 mm) a year of sea level rise. The melt is accelerating rapidly; Antarctica is melting three times faster than it did in the late 1990s and early 2000s, Greenland twice as fast. The logic is simple, though the predictions are complex; uncertainties in timing loom large about how ice loss will or won't accelerate. Ice sheets can become unstable and collapse, rapidly raising sea level. Surging glaciers can stall or move faster. Land can store more or less water. But the overall direction of change is well known: less ice, more ocean.

If all the ice on land melted, it would raise sea level nearly 230 ft. (70 m), but that is highly unlikely in the near term. However, there is a second physics problem increasing sea level and attacking our beaches and coastal cities, one that nearly doubles the effect of melting ice. This second factor is warming (the oceans absorb up to 90 percent of the warming we inflict on the atmosphere, warming we never feel in the air). This is simple science, too—when most stuff heats up, it expands. Think of a hot air balloon. The same physics properties cause heated matter—including water—to expand. Heating the ocean results in sea level rise, even with no additional water added. Currently, this thermal expansion is adding about 0.055 in. (1.4 mm) per year to sea level. That may not seem like much to our limited perception. But consider the energy involved.

Over the period of 1871–2015, greenhouse gasses trapped solar energy and transferred it to the ocean tirelessly—about 43.6×10^{22} joules. A joule is a unit of energy (about the amount of energy you feel when hit by a tennis ball moving at 13.7 mph or 22 km/hr, a pretty soft throw). The amount of solar energy transferred to the ocean over those 144 years is a number so large as to be incomprehensible, and deserves to be written out: 436,000,000,000,000,000,000,000 joules.

The cities most at risk from rising sea levels are also the most populous, as seen here: New York, Miami, Shanghai, and Kolkata, among others. The largest urban areas in the world are located on oceans, exposing billions of people to sea level changes in the next century.

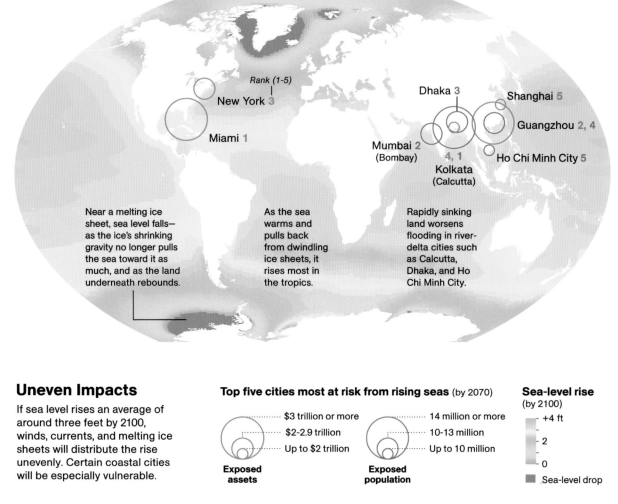

Rank (1-5)

New York 3

Dhaka 3

Shanghai 5

Miami 1

Guangzhou 2, 4

Mumbai 2
(Bombay)

4, 1
Kolkata
(Calcutta)

Ho Chi Minh City 5

Near a melting ice sheet, sea level falls— as the ice's shrinking gravity no longer pulls the sea toward it as much, and as the land underneath rebounds.

As the sea warms and pulls back from dwindling ice sheets, it rises most in the tropics.

Rapidly sinking land worsens flooding in river-delta cities such as Calcutta, Dhaka, and Ho Chi Minh City.

Uneven Impacts

If sea level rises an average of around three feet by 2100, winds, currents, and melting ice sheets will distribute the rise unevenly. Certain coastal cities will be especially vulnerable.

Top five cities most at risk from rising seas (by 2070)

$3 trillion or more
$2-2.9 trillion
Up to $2 trillion

Exposed assets

14 million or more
10-13 million
Up to 10 million

Exposed population

Sea-level rise
(by 2100)

+4 ft
2
0

Sea-level drop

RYAN MORRIS, NGM STAFF. SOURCES: FELIX LANDERER, NASA/JPL; M. PERRETTE ET AL, 2013; ORGANISATION FOR ECONOMIC CO-OPERATION AND DEVELOPMENT

This number beggars analogy, but to make an attempt: the largest explosion most people are familiar with is the Hiroshima atomic bomb, which destroyed an entire city, killed upward of 80,000 people outright, and injured 70,000 more. It released approximately 6.7×10^{13} joules (67,000,000,000,000 joules). The numerical scientific notation commonly used (10^{22} and 10^{13}) tends to obscure the difference in scale. The energy we have injected into the ocean adds up to the equivalent of billions of atomic bomb explosions. And we are adding more now than ever before. Currently, the energy equivalent of 3 to 6 atomic bomb blasts, the kind that leveled Hiroshima, are being added every single second of every single day. Boom, boom, boom.

In the time it took you to read the previous paragraph, around 75 atomic bombs' worth of additional energy went into our oceans in small but unceasing increments and raised the worldwide ocean energy balance and sea level a little bit. In the time it took you to think about that fact, they went up a little more.

The changing seas will take their toll on the social order, in Miami and elsewhere. Globally, current projections have the homes of 100–500 million people underwater by 2100. The most flooded areas will be along the water: oceanfront homes that are soon to be ocean-in homes. (Ironically perhaps, historically poorer neighborhoods set back from the water will remain above the waves, though storm surges and tides will remain a problem for these districts.) The current rate of sea level rise is about 0.15 in. (3.58 mm) per year, and as noted, that rate is accelerating. The Intergovernmental Panel on Climate Change (IPCC) projects rates of 0.6 in. (15 mm) per year in 2100, and nearly 0.8 in. (20 mm) per year after that. While that may not sound like much, it adds up—more than 1.5 ft. (0.5 m) to nearly 2.5 ft. (0.75 m) by 2100. Note that this is a global average and beware the aforementioned problems in conceptualizing scale; some coastal areas will likely see much more. Extreme flood events that used to occur once a century will happen once a year. The numerical rate of change seems slow, but it is inexorable. Get something as large as the ocean moving and it is difficult to stop.

Acidification

We emit a lot of CO_2 into the air, as you are aware. But between 20 and 30 percent of the CO_2 we emit by burning fossil fuels does not stay in the atmosphere very long, it disappears. Where does it go? As you can probably guess from the theme of this chapter, the free-floating carbon dioxide dissolves into the ocean, like carbonating a soda (in fact, that is where the word "carbonating" comes from). It is analogous to kegging your own beer; to make it fizzy, you overload the "keg atmosphere," the head space at the top of the tank, with pure CO_2. It then slowly dissolves into the flat liquid, carbonating it. Uncapping reverses that process, lowering the pressure above the beer and allowing CO_2 to escape. In the case of the oceans, uncapping and lowering the pressure in the head space is not possible, so the carbonation just accumulates.

You might think this is a good thing, that "dilution in solution is the answer to our atmospheric pollution," as some assert. But when CO_2 dissolves into water it does not just disappear. With a little simple chemistry, you can follow this process quite well. (Just recall that elements and atoms cannot typically disappear or change, but they can form different molecules and carry an electrical charge [plus and minus]. Basic structures are built from bonded atoms. It's like building with Legos—the

IN THE TIME IT TOOK YOU TO READ THE PREVIOUS PARAGRAPH, AROUND 75 ATOMIC BOMBS' WORTH OF ADDITIONAL ENERGY WENT INTO OUR OCEANS.

Ocean acidification reduces the ability of many organisms to make shells. This pteropod shell, from a small, drifting sea snail also known as a "sea butterfly," was placed for several weeks in water with carbonate values and a pH similar to that projected for the year 2100. The effects on the shell are obvious.

top left start date March 31, 2007
bottom right end date May 15, 2007

Limacina helicina antarctica - 45 days exposure to seawater simulating the effects of Ocean Acidification
Aragonite saturation projection for the Southern Ocean by 2100 under the IS92a business-as-usual CO2 emissions scenario

pieces can be stuck and unstuck in various ways, but the blocks themselves generally do not change. Each element is abbreviated with a letter or two, like H for hydrogen, O for oxygen, C for carbon, and Ca for calcium.) Water (H_2O; two hydrogen atoms and one oxygen atom) and CO_2 (one carbon and two oxygen atoms) form carbonic acid (H_2CO_3), a weak acid you can easily drink in small quantities (see the aforementioned soda). As with all acids, it likes to release one of those H's (a hydrogen ion, meaning it has one positive charge, denoted H^+) into the solution it finds itself in. We quantify this as a *lowering* of the pH of the solution, making it more acidic and less basic. Lemon juice has a very acidic pH of 2. Bleach is a quite basic pH of 11. The ocean was 8.2, mildly basic (7 is considered neutral). Since the mid-1800s, when coal came into vogue as our first major fossil fuel, the pH of our oceans has dropped to 8.1. That doesn't sound like much, but a 0.1 difference is misleading. Acidity is on a logarithmic scale, which means a pH of 7 is 10 times more acidic than a pH of 8, and 100 times more acidic than a pH of 9. In other words, acidity does not increase linearly with pH dropping; that decline of 0.1 means the ocean is now about 25–30 percent more acidic than it was. By 2100, a decrease of an additional 0.3 pH units is highly likely, which would be more than a 150 percent rise in acidity from historical levels.

So, a 0.1 change is no trivial matter. If your blood pH dropped more than 0.1 pH units, you would enter acidosis and likely die. Marine creatures are in the same boat; adapted to a steady, narrow band of oceanic pH, the acidification of their surrounding environment is similarly problematic. Although difficulties with metabolism, reproduction, and signaling are real, perhaps the most obvious issue is structural. The bodies of many sea creatures, including corals, oysters, many kinds of plankton, and others are constructed largely from calcium carbonate. Vast amounts of life depend on this simple chemical compound, and the organisms that produce calcium carbonate make up the basis of immense ecosystems and food webs. (Recall from the previous section that this built much of Florida!) Calcium carbonate is a straightforward structure: calcium ions (Ca^{2+}) bind with carbonate ions (CO_3^{2-}) to form a powerful little molecule ($CaCO_3$) that can be built into complex shells and protective coatings. Marine organisms do this quite readily. But alas, a problem: carbonate likes to bind with hydrogen ions (those same H^+'s released by carbonic acid), even more than it likes to bind with calcium ions. This forms bicarbonate (HCO_3^-), reducing carbonate ions in the water and thus the ability of organisms to produce those invaluable shells. It sounds a bit complicated, but you're probably familiar with this molecule: baking soda is simply bicarbonate plus a sodium ion [Na^+], forming $NaHCO_3$.

Carbonate floating around the oceans functions as a pH buffer: by the reactions outlined above, it removes hydrogen ions from the water, meaning CO_2 can continue

By 2100, many coral species may not be able to survive due to both heat and acid. While some individual coral species and types can tolerate low pH and warmer temperatures, the future of reef complexes is in doubt.

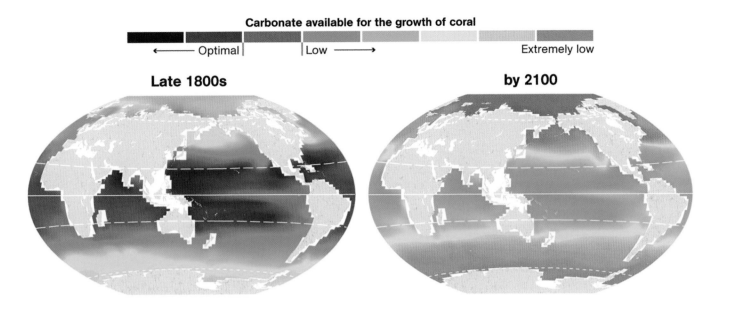

Carbonate available for the growth of coral

←——— Optimal | | Low ———→ | Extremely low

Late 1800s

by 2100

A GROWING PROBLEM FOR CORAL REEFS

In the late 1800s, when fossil-fuel carbon dioxide began to pile up rapidly in the atmosphere and acidify the ocean, tropical corals weren't yet affected. But today carbonate levels have dropped substantially near the poles; by 2100 they may be too low even in the tropics for reefs to survive.

MAPS: TED SICKLEY; NGM MAPS
SARAH COOLEY, WOODS HOLE OCEANOGRAPHIC INSTITUTION (MAPS)

to dissolve at a relatively high rate while acidity increases only modestly. But doing so makes carbonate harder to come by for those marine organisms. They can respond by expending more energy to concentrate calcium and carbonate together, still building their shells and structures, but at increased energy costs. Some species, like lobster, respond by building larger and thicker shells. Other populations, like some corals and clams, wither. And while the response to ocean acidification may vary among its inhabitants, the result is a destabilized ecosystem as winners and losers are increasingly chosen by the chemistry of the ocean.

FRESH WATER

Earth is an ocean planet, a saltwater realm. But terrestrial life is dependent on fresh water, though it constitutes only 2.5 percent of all the planet's water (and only about one percent is non-frozen and relatively accessible). We and our fellow land life-forms rely on that tiny and ephemeral portion of the hydrosphere. The saltwater world is slow, ponderous; the freshwater world is quick and rapid, bouncing from surplus to deficit.

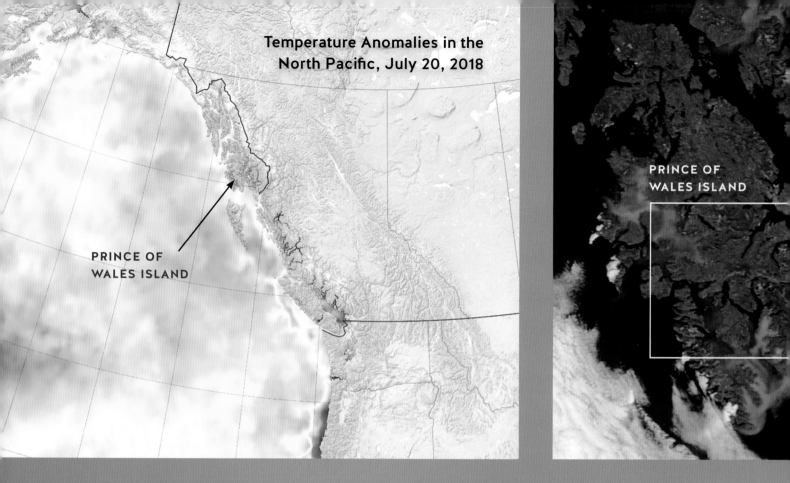

Temperature Anomalies in the North Pacific, July 20, 2018

PRINCE OF
WALES ISLAND

PRINCE OF
WALES ISLAND

Blooming Algae

The ongoing journey of oceanic change involves drivers and passengers. The drivers are things like warming temperatures or CO_2 concentrations; passengers are things like sea otters, kelp, and lobsters. The problem with simple analogies is that there are a lot of both drivers and passengers, and they often switch places in the car. Moreover, each influences the other, changing impacts. Currents change. The ocean gets less salty, or more. The knock-on effects are endless. A symptom of this chaos is the emerging frequency of algal blooms, so large they are visible from space. These blooms are not a wholly novel phenomena, but the frequency and scope of their emergence today is unprecedented. In the above images, algal blooms seen from space stand out as light-colored patches of water against the dark blue coastal Pacific.

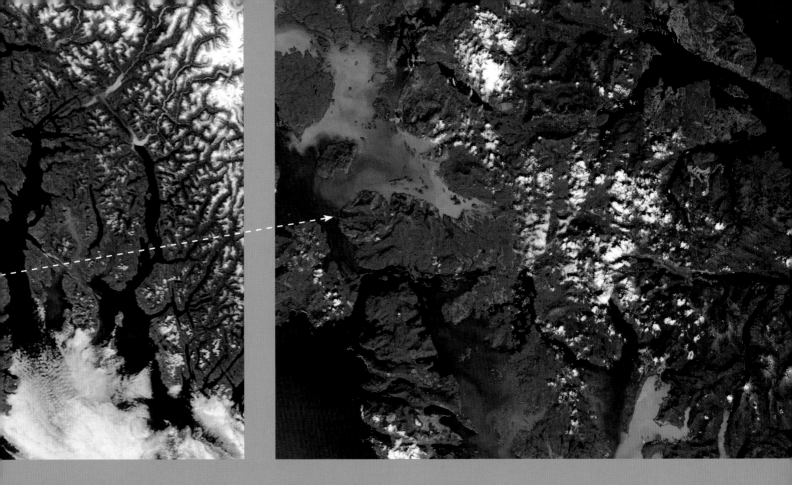

In the waters off Prince of Wales Island, in southeast Alaska, whole fjords can fill with a milky white layer of what is probably the frothy algae *Emiliania huxleyi*. This particular "coccolithophore" (a tiny, photosynthetic plankton) is a worldly one, found nearly everywhere and frequently forming the founding layer of complex marine food webs. *Emiliania huxleyi* is sensitive to temperature, and particularly loves warm, low-nutrient environments—traces of *E. huxleyi* are used to reconstruct oceanic conditions going back thousands of years. *Emiliania huxleyi* is primed to take advantage of the abnormally hot, unpleasant-for-typical-local-life waters of the new millennium. Like weeds taking over a garden, there are winners and losers with oceanic climate change, organisms that can tolerate the new reality and run wild with little competition.

NORTH AMERICA.

#		Feet
1	Popocatipetl	Feet 17,735
2	Orizaba	17,368
3	Iztaccihuatl	15,700
4	Long's Peak	15,000
5	Oregon Mountains	13,000
6	Mt St Elias	12,680
7	James Peak	12,000
8	Volcano of Colima	9,186
9	Mt Fairweather	8,940
10	City of Toluca	8,818
11	Town of Isla Huaca	8,481
12	Town of Perote	7,723
13	City of Mexico	7,410
14	City of Puebla	7,198
15	City of Durango	6,841
16	White Hills N.H.	6,634
17	City of Valladolid	6,404
18	Source of Missouri Riv.	5,000
19	Mt Hecla	5,000
20	Peaks of Otter	4,260
21	Killington Peak	4,000
22	Table Mountain	4,000
23	Catskill Mountains	3,804
24	Pine Orchard House	3,104
25	Watchusett Mt	2,990
26	Alleghany Mountains	2,400
27	Butler Hill	1,535
28	Source of Mississippi R.	1,293
29	Pittsfield Mass.	1,000
30	Frankstown Penn.	910
31	Pittsburg Penn.	678
32	Lake Superior	641
33	Pottsville Penn.	620
34	Buffalo N. York	578
35	Cumberland Maryland	573
36	Lake Erie	568
37	Rochester N. York	506
38	Lebanon Penn.	480
39	Worcester Mass.	450
40	Utica N. York	425
41	Harrisburg Penn.	300
42	Lake Ontario	231
43	Shot Tower at Philad.	184
44	Easton Penn.	161
45	Washington City	60

SOUTH AMERICA.

#		Feet
46	Mount Sorato	Feet 25,250
47	Mount Illimani	24,350
48	Chimborazo	21,730
49	Disco Casada	20,892
50	Corcobado	20,000
51	Cayambe Urcu	19,386
52	Antisana	19,400
53	Tajora	19,200
54	Cotopaxi	19,000
55	Volcano of Arequipa	18,400
56	Illinissi	17,238
57	Sangu	17,136
58	Tunguragua	16,500
59	Cerro de Potosi	16,037
60	Pichincha	16,000
61	Mines of Potosi	15,912
62	Curguirazo	15,540
63	Nevada de Merida	15,201
64	Village of Tacorra	14,250
65	City of Potosi	13,668
66	Gap of Huessos	13,605
67	Mines of Huancavillica	13,600
68	Plains of Assuay	13,125
69	City of Puno	12,832
70	Tiaguanaco	12,812
71	Lake Titicaca	12,703
72	La Paz	12,195
73	Gaharapata	11,641
74	Mines of Choto	11,562
75	City of Tupisa	10,000
76	City of Quito	9,540
77	City of Chuquisaca	9,331
78	City of Bogota	8,818

HEIGHTS OF THE PRINCIPAL MOUNTAINS IN THE...

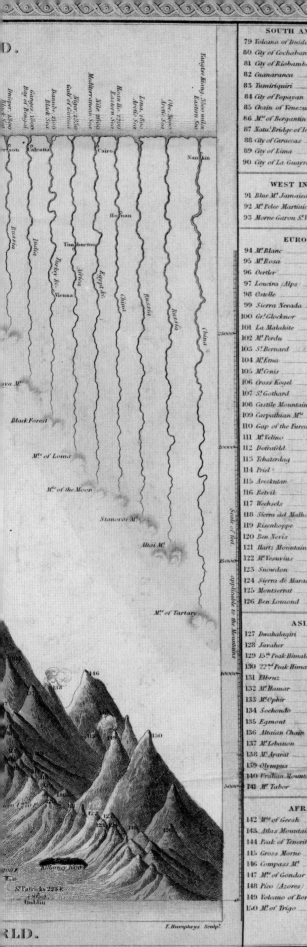

SOUTH AMERICA.

79	Volcano of Duido	Feet 8,467
80	City of Cochabamba	8,448
81	City of Riobamba	8,441
82	Guanaranca	6,420
83	Tumiriquiri	6,250
84	City of Popayan	5,825
85	Chain of Venezuela	5,000
86	M.ⁿ of Bergantin	4,500
87	Natu.l Bridge of Icononzo	2,930
88	City of Caraccas	2,860
89	City of Lima	512
90	City of La Guayra	20

WEST INDIES.

91	Blue M.t Jamaica	Feet 7,431
92	M.t Pelee Martinico	5,100
93	Morne Garou S.t Vincent	5,050

EUROPE.

94	M.t Blanc	Feet 15,668
95	M.t Rosa	15,552
96	Oertler	15,430
97	Loucira (Alps)	14,451
98	Ostelle	12,800
99	Sierra Nevada	12,762
100	Gr.t Glocknor	12,714
101	La Malahite	11,421
102	M.t Perdu	11,283
103	S.t Bernard	11,011
104	M.t Etna	10,780
105	M.t Cenis	9,956
106	Gross Kogel	9,700
107	S.t Gothard	9,075
108	Castile Mountains	9,000
109	Carpathian M.ts	8,640
110	Gap of the Furca	8,301
111	M.t Velino	8,000
112	Dofrafeld	7,620
113	Tchaterdag	6,600
114	Priel	6,565
115	Areskntan	6,180
116	Retvik	6,000
117	Wechsels	5,686
118	Sierra del Malhao	5,500
119	Risenkoppe	5,070
120	Ben Nevis	4,379
121	Hartz Mountains	3,926
122	M.t Vesuvius	3,932
123	Snowdon	3,568
124	Sierra de Marao	3,500
125	Montserrat	3,300
126	Ben Lomond	3,262

ASIA.

127	Dwahalagiri	Feet 26,262
128	Javaher	25,745
129	15th Peak Himalaya	21,617
130	22nd Peak Himalaya	19,497
131	Elbruz	18,000
132	M.t Hamar	15,000
133	M.t Ophir	13,842
134	Sochondo	12,800
135	Egmont	11,433
136	Altaian Chain	10,500
137	M.t Lebanon	11,000
138	M.t Ararat	9,500
139	Olympus	9,400
140	Uralian Mountains	4,700
141	M.t Tabor	2,000

AFRICA.

142	M.ts of Geesh	Feet 15,000
143	Atlas Mountains	12,500
144	Peak of Teneriffe	12,176
145	Gross Morne	10,000
146	Compass M.t	10,000
147	M.ts of Gondar	8,500
148	Pico (Azores)	7,016
149	Volcano of Bornou	7,600
150	M.t of Trigo	7,400

Rivers and Mountains of the World

Rivers and mountains both carry immense significance for human societies. The beauty of both, plus significant amounts of scientific data, are included on this 1849 chart by Samuel Augustus Mitchell. While the mountains that power the rivers take dramatic center stage here (though note the absence of Mount Everest and other peaks in Asia, which were not accurately measured until 1856, and Denali in North America, similarly uncharted at the time), the complex river courses and the societies they nurtured—extended linearly to show their mouths, lakes, headwaters, and complex tributaries while also comparing their relative lengths—are perhaps more significant.

Human civilization started on rivers. From the Tigris and Euphrates of biblical fame to the Yangtze that nourishes China, freshwater rivers inevitably populate the origin myths of humankind. And with good reason—rivers provide an immense amount of resources. They are a constant input of fresh water to a growing city. They are a source of sanitation, carrying waste away. They provide a moving conveyer belt for transportation of resources from upstream, and a way to efficiently move goods and products to trading partners. They are a source of mechanical energy. Strategic locations, like the mouths of major watersheds, can effectively integrate the entire catchment in terms of political, economic, and ecological power. The magnetism of rivers is evident in human society up and through the Industrial Revolution, as people continuously moved to communities located on their banks—cities such as London, Paris, and New York. That trend has somewhat reversed in the post-industrialization era; people have recently (in the 20th century) begun migrating to dry landscapes for the sun, exploiting groundwater to survive. But while technology has somewhat liberated us from the need for riverine resources (water, sanitation, transportation), the movement of people into arid regions only exacerbates the eternal problem of fresh water. Regarding water availability specifically, this dryland trend is made possible by two technologies. The first, a reliance on groundwater, is a temporary solution to the ongoing need for water and will be discussed later. The second is our relatively recent ability to carve new waterways, diversions, ditches, and tunnels to deliver water where we want it. Instead of humans moving to the water, we now can move the water to us in a steady flow of fresh supplies. Humans now make the rivers. It is more predictable, reliable, and… "civilized." But doing so serves to remove ourselves from the natural rhythms of the world, beats and pulses that define a dynamic and living system. And it is only temporary.

A living river

All those diversions and ditches and regularly delivered water supplies, the reliability of your tap, make it quite easy to forget one vital fact about natural rivers. They vary dramatically, from hour to hour, day to day, season to season, and century to century. In the mountains, early morning sees stream channels nearly dry as the night's chill locks down the water in snow and ice. In the afternoon, those dry channels are burbling merrily with the warming flow. Over a season, rivers wax and wane as yearly temperature and precipitation changes make their mark; in high-latitude locales, snowmelt can swell a river for a few months (termed a "freshet"), then that river may drop to nearly dry later in the year. In the tropics, dry seasons and wet seasons can literally make rivers and river deltas appear and disappear annually.

The world can be seen purely as watersheds, draining to the sea or to isolated inland basins. Here, each color represents a North American drainage basin from source to sea, mountains to ocean, a unique water-defined landscape that integrates all the land by its connection to water. Each is a unit unto itself, delineated by topography and united by the flow of water. They integrate the terrestrial landscape into singular points such as the Mississippi watershed that neatly covers the interior of North America (in deep blue). This is the world the rains experience on their journey from the clouds to the sea.

Meandering Mississippi

Harold Fisk and the US Army Corps of Engineers worked over 2000 miles of river for 3 years to resurrect the historical meanderings of the greatest river in North America, the Mississippi. Using a combination of historical maps, old roads, geological techniques, labor-intensive hole digging, and basic intuition, their project, shown in part here, beautifully illustrates that even the largest rivers can move, and frequently do. The entire floodplain is a riot of color, indicating far-flung meanderings over time (legend on page 90; each color is a different former river channel, from a time when the river took a slightly different path through the floodplain). In Plate 22, Sheet 11, in the far upper right corner, the town of Vicksburg, Mississippi, can be seen perched on the edge of the historical river. Vicksburg was a bustling river port when it was founded in 1811; then the river snaked up north and down south around the Desoto Peninsula, visible as the blue and red channels (the river course as of 1765 and 1820, respectively). But on April 26, 1876, the river breached that spit of land. Vicksburg was without its riverfront (visible as the new 1880 channel, in green). The town would have shriveled as well, but a multidecadal river diversion project, which changed the course of the Yazoo River to flow past Vicksburg's former riverfront and into the Mississippi, restored Vicksburg's river access. The city was saved. But it remains a dangerous victory. Flooding is commonplace and while Vicksburg sits at a relative high point of the river, major floods are always a danger. Charlie Patton immortalized the scene in Vicksburg during the Great Flood of 1927—and the human struggle to find dry ground somewhere, anywhere—in the foundational blues song, "High Water Everywhere." And indeed, a living river can be an unpredictable thing, wrapping and twisting and making dry land hard to find. Page 90 shows another part of the river upstream, near New Madrid, Missouri.

PLATE 22
SHEET 11

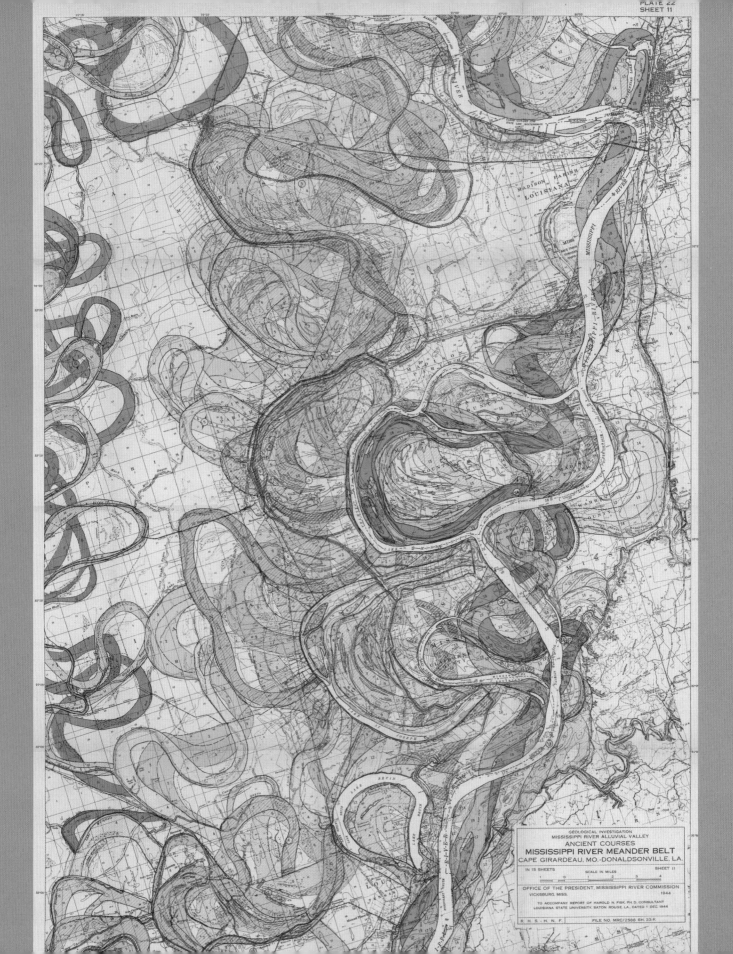

GEOLOGICAL INVESTIGATION
MISSISSIPPI RIVER ALLUVIAL VALLEY
ANCIENT COURSES
MISSISSIPPI RIVER MEANDER BELT
CAPE GIRARDEAU, MO.-DONALDSONVILLE, LA.

IN 15 SHEETS   SCALE IN MILES   SHEET 11

OFFICE OF THE PRESIDENT, MISSISSIPPI RIVER COMMISSION
VICKSBURG, MISS.   1944

TO ACCOMPANY REPORT OF HAROLD N. FISK, PH.D. CONSULTANT
LOUISIANA STATE UNIVERSITY, BATON ROUGE, LA, DATED 1 DEC. 1944

R. H. S. - H. N. F.   FILE NO. MRC/2586 SH. 33-K

PLATE 22
SHEET 2

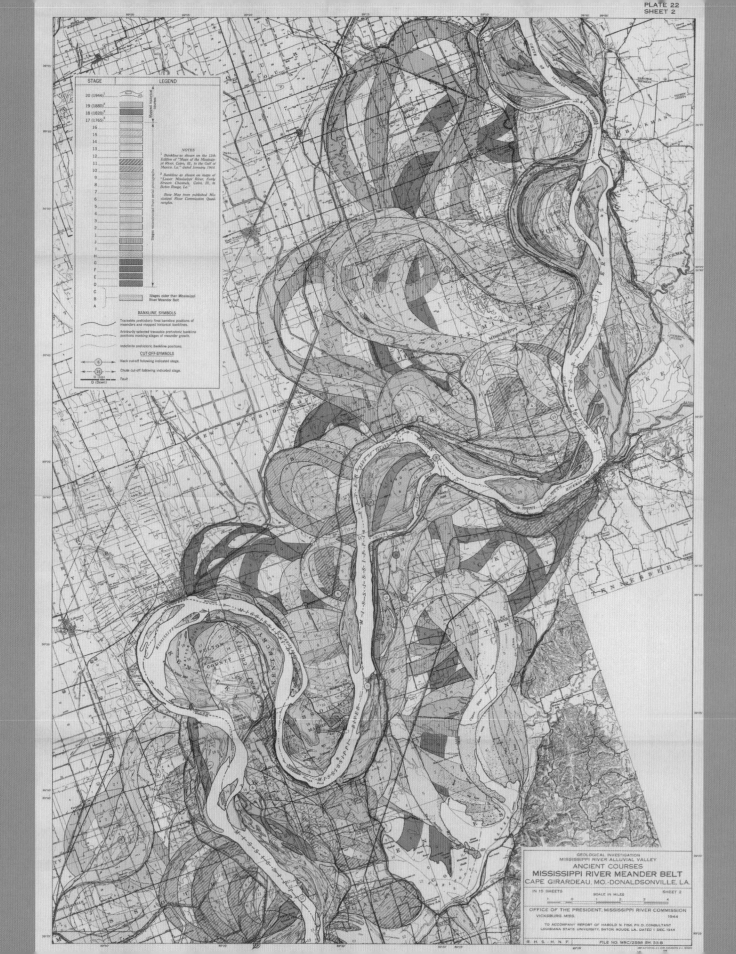

STAGE | LEGEND

20 (1944)
19 (1880)
18 (1820)
17 (1765)
16
15
14
13
12
11
10
9
8
7
6
5
4
3
2
1
J
I
H
G
F
E
D
C
B
A

Stages older than Mississippi
River Meander Belt

BANKLINE SYMBOLS

CUT-OFF SYMBOLS

Fault

GEOLOGICAL INVESTIGATION
MISSISSIPPI RIVER ALLUVIAL VALLEY

ANCIENT COURSES
MISSISSIPPI RIVER MEANDER BELT
CAPE GIRARDEAU, MO.-DONALDSONVILLE, LA.

IN 15 SHEETS   SHEET 2
SCALE IN MILES

OFFICE OF THE PRESIDENT, MISSISSIPPI RIVER COMMISSION
VICKSBURG, MISS.
1944

TO ACCOMPANY REPORT OF HAROLD N. FISK, PH.D. CONSULTANT
LOUISIANA STATE UNIVERSITY, BATON ROUGE, LA. DATED 1 DEC. 1944

Tracking and predicting this variability was a matter of life or death for the civilizations on the rivers' banks. The ancient Egyptians created the first flood gauges, "nilometers," to record this annual change on the Nile River. The simplest version is a series of steps that walk down into the river, from which priests (the earliest river scientists) could watch for fine-scale variability in the river height and then predict the annual flood, which replenished the floodplain and allowed Egypt to grow enough food to feed the largest empire of its age. While thousands of years separate us from those early scientists, don't be fooled into thinking modern technology protects us from nature's fluctuations—the empires of today are just as sensitive to their rivers as the ancient Egyptians.

In 1944, Harold Fisk and the United States Army Corps of Engineers took on the task of quantifying the historical Mississippi. It is a monstrous natural wonder, a blue line bisecting much of North America. It defines the American Midwest and South. And like a slowly writhing snake, the Mississippi—at 3780 miles (6084 km) the 4th largest river in the world—contorts and flexes powerfully as it flows south. Fisk's goal was to chart and define the landscape that the river claimed as it moved from near the Great Lakes to the Gulf of Mexico. The river you see today is only the current incarnation. In fact, the river proper, including all its associated bottomlands, is much larger. Today's snapshot in time is only part of the story.

THE RIVER YOU SEE TODAY IS ONLY THE CURRENT INCARNATION.

The mapping effort by Fisk and company was substantial. Historical records served for a few hundred years of the most recent past, along with traces imprinted on the landscape, such as roads that curved around now-filled-in oxbow lakes and former river towns abandoned high and dry on an old bend. Bridges and ferries attested to the river's wandering history with their built-by and abandonment dates. For channels before that, Fisk relied on the geological tenet that younger material is laid down on top of older material (the "principle of superposition"). In other words, by digging down, Fisk could chart former channels buried under newer ones. The result is a multifaceted ribbon of meanders, twists, and turns morphing into oxbow lakes, like watery commas, which eventually fill and are overgrown—only to be reclaimed and washed away a few hundred years later as the river reasserts its dominance.

Rivers are formed by topography. In steeper terrain, a river's path is set primarily by things like geology—where the cuts and dips of the rocks themselves channel waterflow. Over time, the rivers cut down into the ground, reinforcing and solidifying their paths and occasionally forming canyons and slots. In these systems, the

Sauk River Via Satellite

The Sauk starts in the beautiful North Cascades of Washington State, a sharp series of peaks that rival any mountain range in the world for beauty. Its name derives from the Sah-kee-ma-hu band of the Skagit people, who have inhabited this landscape from mountains to sea for millennia. From above, the river braids through an active gravel flat, scoured each year by ample winter rains and home to salmon, trout, and the occasional angler. But it appears tightly bound by vegetation, the glorious evergreens, cottonwoods, and vine maples of the Pacific Northwest. The human presence is also unmistakable. Farm fields take advantage of the flat land near the river and are graced by curving rows of trees (logging cuts are readily apparent). Houses and roads parallel the river course, enjoying the beautiful stretch of land overlooking the river. One can almost hear the sound of fast water sliding over the smooth stones, season after season, year after year.

Sauk River Via LiDAR

Rivers such as the Sauk reveal much more when seen through the lens of modern methods. In this LiDAR image, which shows the same segment of the Sauk pictured previously, we can see the broad floodplain, the same dynamism and beauty, that Fisk charted in 1944 on the Mississippi—all encoded as subtle variations in elevation.

The past imprint of the twisting river remains today on the landscape as almost invisible dips, channels, and banks, sometimes only centimeters from the surrounding land. In this LiDAR imagery, white shows the lower areas (where the river is now), fading to slightly higher elevations in darker blues. Former oxbows are faintly impressed on the background, like footprints off the main channel; old dark riverbanks pop out like backstops. These maps demonstrate that the riverine system is alive horizontally as much as vertically, and firmly put modern infrastructure, like the houses, roads, and fields seen previously, in their place in the dynamic natural system.

variability is not so much spatial as temporal: flash floods or seasonal pulses like those that characterized the Nile. The Grand Canyon is in the same place it was millions of years ago. It hasn't moved much. But as the gradient mellows out and the rivers slow, the variability reverses and becomes as much spatial as temporal. The rivers, freed from the constraining walls of rock and breathing a bit deeper, can meander. They can push out against their banks, and those often move. The interplay between the river and its landscape becomes one of give and take, with the occasional pulse of violence supplied by floodwaters.

First, the give and take. Quite literally, rivers sculpt their own landscapes. In high-gradient streams it is primarily a one-sided relationship, with waters excavating down and carrying sediment away like a conveyer belt. But the Mississippi, and similar flatwater rivers, also give land—they are as much bulldozers as excavators. The Mississippi River alone carries over 606 million short tons (550 million metric tons) of sediment into the Gulf of Mexico each year (hence the nickname "The Big Muddy"); the Amazon River delivers about 1.1 billion short tons (1 billion metric tons) to the Atlantic. That sediment, be it delivered from steeper topography upstream or from the former channels on-site, swirls and tries to settle constantly. Quick-moving water keeps picking up the soil, but if the water slows, the dirt settles out of suspension. As a result, the river dynamically slows itself—at every curve, the inside part of the river slows down and the sediment falls to the bottom, infilling the current channel. The outer part of the curve maintains speed as water sloshes around the bend, eroding the opposite bank. This pushes the water into a larger turn, stretching out horizontally like a lazily stretching snake, which slows the river even further by compressing more river-miles into the same vertical drop (after all, water flows downhill, but there is only so much vertical to work with. The more river that is packed into a vertical area, the gentler the slope that river experiences).

But as the coils pile up, the corners begin to erode toward each other. This is when the violence happens. One particularly snowy winter is followed by a rainy spring, melting fields in days and causing a massive pulse of water. Or a hurricane slams into the area. The banks overflow, and the coils of the river connect. Suddenly there is a shorter route downhill. Rapidly, the river snaps back to a straighter configuration and the former channel is left as an oxbow lake, which slowly fills in over the years until all that is left is a curlicue slip of abnormally flat land in the floodplain. However, even the new situation is only a temporary interlude—the river continues

THE INTERPLAY BETWEEN THE RIVER AND ITS LANDSCAPE BECOMES ONE OF GIVE AND TAKE.

to carry sediment, and little variations in flow excavate and drop sediment into new configurations that create future convolutions and oxbows.

This variability approaches the scale of human experience, as Fisk's survey shows. Sometimes it overwhelms it. Dynamic flooding and bank-busting events are not unusual, but rather are simple reassertions of the river's rights to its territory, the broad floodplain which, while not always wet, is as much a part of the broader riverine system as the waters themselves.

Despite human desire for stability, input variability—more rain one year, more snow another—and the erosive and depositional nature of rivers themselves mean dynamism. For species on the rivers' banks, the ground is constantly shifting beneath their feet as waterways abandon and then reclaim the land of their creation.

Variability to the extreme

The rush of seasonal snowmelt filling dry arroyos and the charts of past river channels testify that our freshwater resources are variable. Some of that variability is self-induced, the ponderous interplay between tons of sediment lazily settling in still water and the river coiling and uncoiling in its consistently shifting banks. But much is also weather driven.

As the climate warms, the atmosphere can carry more water. Globally, net precipitation should increase simply due to physics. But along with more water, we are also seeing increases in variability: timing, location, and amount. And although the planet is clearly too large to discuss every nuance, the major culprits in this increased variability can be broken into two categories: "pulse" disturbances—things like hurricanes, typhoons, and other short-timescale disruptions, and more gradual "push" activities—long-term fluctuations that change over the timeline of a year or more. The likelihood of both types of circumstances is being affected by global warming. (Ascribing the occurrence of any single event to climate change is the wrong way to think about things. It's more properly conceptualized as changing probabilities of extreme events along with the steady drumbeat of changing averages.)

First, the pulses. Storms, whether hurricanes, typhoons, thunderstorms, or blizzards, deliver an immense amount of fresh water at one time. If more water is delivered than can be either stored in the soil or exported via drainages, the result is flooding. And, again, while attributing any individual storm to climate change is notoriously difficult, it is straightforward to realize that (1) increased human population densities result in increased monetary-infrastructure-human damages from any given storm, and (2) warmer air holds more water, meaning storms can pack a larger punch.

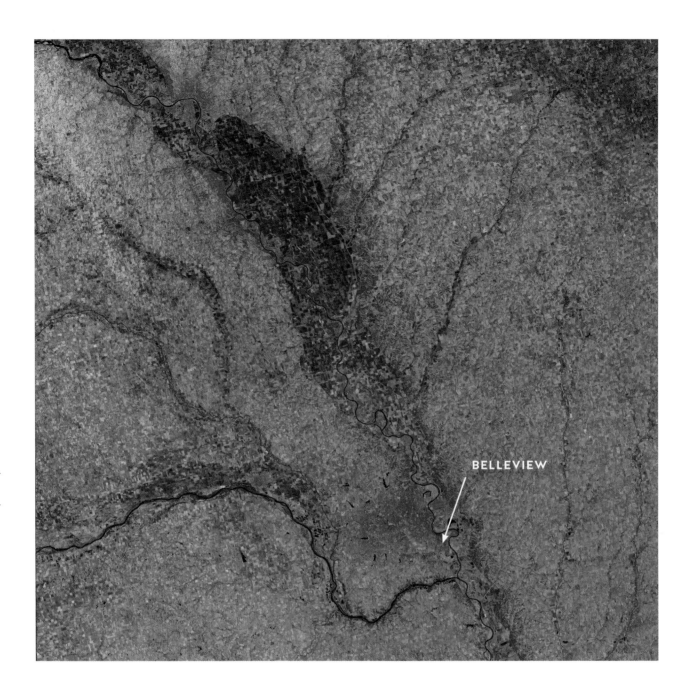

BELLEVIEW

In March 2019, the Missouri River reasserted its claim to a wide swath of its floodplain, a result of frozen ground, ice-choked water, and late winter rains. The oxbow bend looping out from Belleview, Nebraska, was temporarily reclaimed as levees failed in some locations. These pre- and during-flood images show how the flood is merely a filling of available topography, a predictable result of natural variability at the scale of a floodplain.

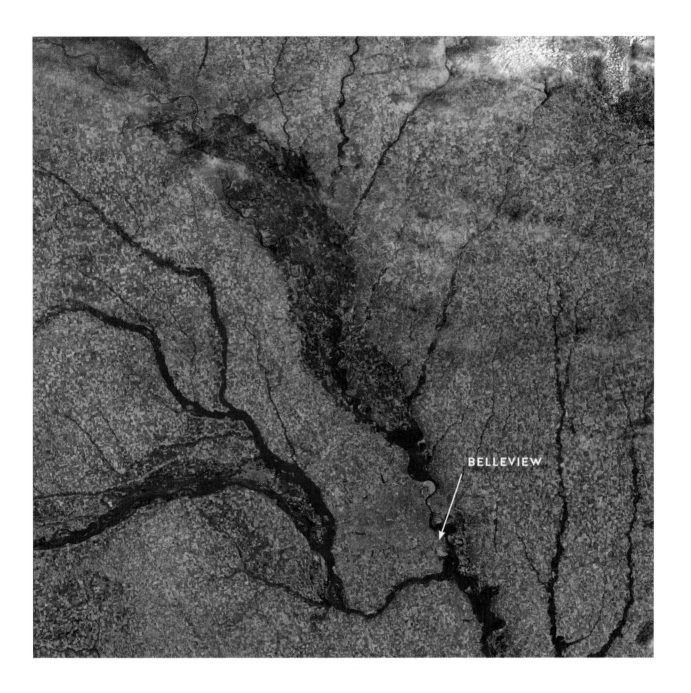

BELLEVIEW

Lake Bonneville

Over millennia, variability in precipitation can reshape entire regions. The US interior Southwest, a land of deserts and scrub, was once dominated by Lake Bonneville. Covering most of northwest Utah and parts of Nevada and Idaho, the lake rose and fell gently with prehistoric climate change—sometimes flooding the basin until mountains were little islands, other times contracting. The lake was around 1000 ft. (300 m) deep at its greatest extent. Like rivers, lakes shape their landscape. Lake Bonneville rose and fell, creating terraces and paleo-shorelines which can be found throughout Utah today, including terraces and ledges above Salt Lake City, itself built on the historic lake bottom. This is beautifully illustrated by Grove Gilbert, who pieced together the shorelines in the 1890 map shown here, which includes the Idaho border on the northern edge and the current distribution of the Great Salt Lake in light gray-blue, clearly only a remnant of the former glory of Bonneville. Around 14,500 years ago, the lake crested at the same height as Red Rock Pass on the northern edge, just into what is now Idaho, and over which waited the Snake River and the Pacific Ocean, downstream. A massive collapse of the ridge at that level resulted in one of the largest floods in recent geological history, nearly 35 million cu. ft. (one million cu m) per second of water pouring down into the Snake River and significantly altering the geography of the region. The water level stabilized nearly 330 ft. (100 m) lower. That newer level created the "Provo shoreline," which is still easily visible around Utah. Since that time, what was left of Lake Bonneville has continued to shrink due to evaporation (and scant inputs) as the climate dried.

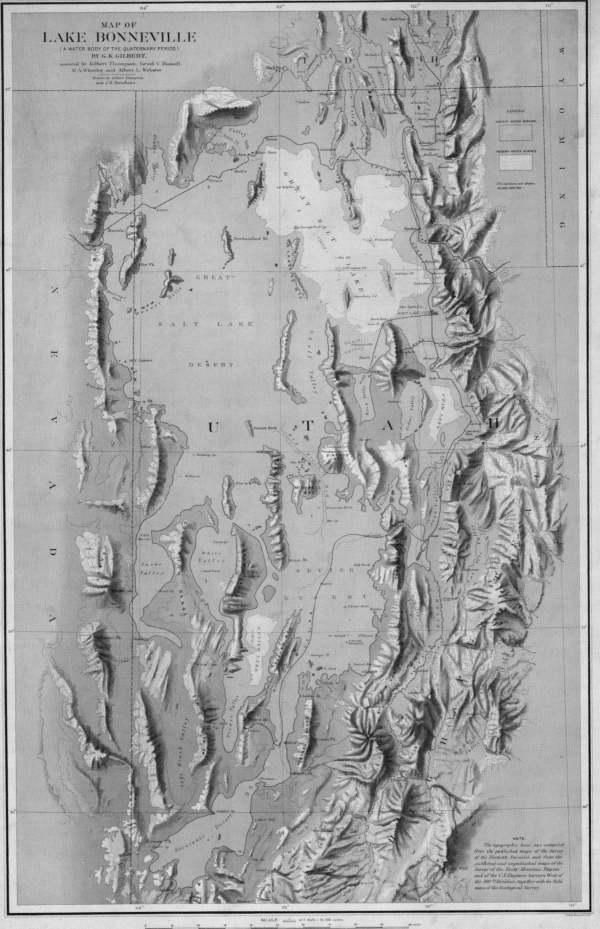

MAP OF
LAKE BONNEVILLE
(A WATER BODY OF THE QUATERNARY PERIOD)
BY G.K.GILBERT,
assisted by Gilbert Thompson, Israel C. Russell,
H.A.Wheeler and Albert L. Webster.
Drawn by Gilbert Thompson
and J.H. Renshawe.

LEGEND.
ANCIENT WATER SURFACE.

MODERN WATER SURFACE.

The contours are drawn
at each 500 feet.

NOTE.
The topographic base was compiled
from the published maps of the Survey
of the Fortieth Parallel, and from the
published and unpublished maps of the
Survey of the Rocky Mountain Region
and of the U.S. Engineer Surveys West of
the 100° Meridian, together with the field
notes of the Geological Survey.

SCALE

Satellite imagery has revealed the power of these pulsed storm events in dramatic detail. And although there are a host of intense weather patterns, two worth mentioning in more detail are atmospheric rivers and hurricanes (again—they are impressive processes!). Both transport an immense amount of fresh water long distances, both can result in flooding, and we struggle to predict the future trends in both—they represent significant unknowns to our freshwater knowledge.

Picture a river roughly 300 mi. (500 km) wide, as wide as the state of Pennsylvania or as far across as the Netherlands and Belgium combined. Now imagine that river stretching the length of the entire Pacific Ocean and carrying around 15 times the flow of the Mississippi River at about 30 mph (50 km/hr). Now picture it in the sky. These are atmospheric rivers, and they dwarf terrestrial ones. Atmospheric rivers are naturally occurring phenomena that form as the twists and turns of high- and low-pressure systems in the atmosphere trap and focus tropical water vapor into narrow corridors of moisture, which then flow rapidly from wet tropics to higher latitudes. When they hit mountains, some of the most intense rains on the planet result. Asia and the high Himalaya are beset by atmospheric rivers, which can cause massive landsliding and floods. California receives up to half of its annual precipitation from these rivers, which slam into the mountains and dissipate in just a few days each. This underscores the dual nature of freshwater resources: without these

An atmospheric river snakes toward California, its origins near Taiwan.

MAP: JEROME N. COOKSON, NGM STAFF. SOURCE: NOAA

Dry air   Moist air

massive pulses—variability our infrastructure struggles to harness—the agricultural capital that is California would not grow. With climate change, this precious yet dangerous catastrophe is moving north, drying out the lower latitudes of California and drenching locations higher on the West Coast of the United States.

On to hurricanes. Tropical storms and hurricanes are enormous heat engines, pumping energy from the water into the atmosphere. They too are sources of significant freshwater inputs and enormous variability. In 1979, Tropical Storm Claudette produced 43 in. (109 cm) of rain in Alvin, Texas, over 24 hours. This US record stood for decades. In 2018, Hanalei (on the north shore of Kauai, a beautiful beach town I briefly lived near as a conservationist after college) broke the record with just under 50 in. (126 cm) in 24 hours. That is substantially more than Seattle, Washington, receives in an entire year. In August 2017, Hurricane Harvey delivered 60 in. (150 cm) of rain to Nederland, Texas, over multiple days. These are immense pulses of water. New satellite imagery allows us to visualize at a very high resolution the variation in rainfall rates and amounts that would have previously been impossible, and to chart changes in precipitation over time.

Although it is not yet clear if hurricanes are increasing in frequency, they do seem to be getting wetter. To stick with Texas, once-every-100-years rainfall events (often associated with tropical storms or hurricanes) are now once-every-25-years events. For every 1.8°F (1°C) of atmospheric warming, there is an additional 7 percent of moisture in the air. Air at 68°F (20°C) holds twice as much water vapor as a cooler 50°F (10°C) breeze. That's a double whammy for tropical storms. They are stronger if they do occur (warmer water, more heat to pump) and can thus pull in moisture from a larger area, plus they can hold more water aloft across that area. So any debate about hurricane frequency and climate change is not really the whole story; regardless of changes to frequency, we will see changes to hurricane impact. The pulses of fresh water are both life giving and life taking, the more so as we require more water for agriculture and build more and more cities along rivers and coasts for our growing populations.

Those are the pulse events: dramatic, short-term water dumpers. What about the push activities, bigger atmospheric patterns that result in persistently wet seasons and years? Those can have major impacts too, worse than any hurricane because they tend to occur on broader scales—regions, not counties.

The foot-stomping, reverb-soaked drum line of Led Zeppelin's "When the Levee Breaks" pounds out a true story. Originally recorded as a blues song by Memphis Minnie and Kansas Joe in 1929, the song was inspired by the 1927 Great Mississippi

ALTHOUGH IT IS NOT YET CLEAR IF HURRICANES ARE INCREASING IN FREQUENCY, THEY DO SEEM TO BE GETTING WETTER.

Flood, the most destructive flood event in US history (costing $1 billion, equivalent to one-third of the *entire* US federal budget at the time). But it was not a single storm that drove this particular event. There's a common misconception that drainages (from creeks to storm drains) pull off water as fast as it rains, preventing flooding. That's not generally true. The reason most storms do not cause flooding is because the ground soaks up a lot of water, trapping it around soil grains and in pores between those grains. Then it slowly drains over a few days or weeks. Water levels in rivers do not rise substantially until the ground is saturated, except in the heaviest rainstorms, when water flows over the ground. A very wet late summer in 1926 had left the entire US Midwest's soil saturated, with local flooding in September and October with every little shower. One observer wrote:

> The foundation was thus so well laid that neither prophetic vision nor vivid imagination was required to picture a great flood in the following spring, contingent only upon a rainfall substantially above average during the winter months.
> —Alfred Henry, *Monthly Weather Review*, 1927

They got that rainfall. A series of storms in March and April 1927 falling on the already primed, saturated soils produced record flooding over 27,000 sq. mi. (70,000 sq km). It was not that each individual storm was record breaking, though they were intense, but that the clustering of storms in space and time, coupled with that recent history of wet soils, led to oversaturation and floods. Seven hundred thousand people lost their homes. Over 500 died. Floodwaters spanned 10 states, hitting most dramatically in Arkansas, Tennessee, Mississippi, and Louisiana. The economy collapsed, driving mass migrations of humanity from the flooded South to the urban North, inspiring generations of artists and songs that are still sung today.

Science is fairly confident that floods are becoming more likely in many locations today, as are their arid companions, droughts. The snowmelt season in high latitudes is compressing as winters warm and rain now often falls instead of snow. "Rain on snow" events cause rapid melting, which can choke rivers with the water they historically would have drained over a few months. This trend is expected to increase with time. Warmer, wetter weather. Storms. Faster melt. The capricious nature of river and fresh water defy our attempts to control them. We build levees, and they can break.

All the variability that is wrapped up into something as dynamic as the natural freshwater cycle has been tamed for large civilizations to persist. And when thinking

In Texas, water arrived in buckets from June 18–21, 2018. Nearly 20 in. (500 mm) of rain fell in just 72 hours in Corpus Christi alone (on the southern Texas coast). Other areas saw more; the white areas on this map show the locations of the most intense precipitation as the storm approached. Rivers carried 400 times their normal flows. Houses and cars were completely submerged. All from a slow-moving low-pressure system that never even reached tropical storm status but remained a series of thunderstorms and rain.

about the natural world of water, we cannot ignore the effort that one industrious species, *Homo sapiens*, has put into engineering the variability *out* of the natural world. As a species, we like stability and predictability, not change and unpredictability. But as stated previously, water is complex to manage. Water does not compress for easy stockpiling, so it is hard to store large quantities in one place. It tends to evaporate if left in the sun too long. It can get contaminated. It is heavy, and hard to move long distances. The distribution of fresh water is highly variable; some places have water in surplus, other places have very little. And we need a lot of it. The average person uses 25–100 gal. (100–400 l) of water in their home every day, and 10 times that if you account for irrigation used to produce food eaten throughout that day. It is not inexhaustible, or replaceable, or manufacturable. As *National Geographic* said so succinctly in 1993, "All the water that ever will be is, right now."

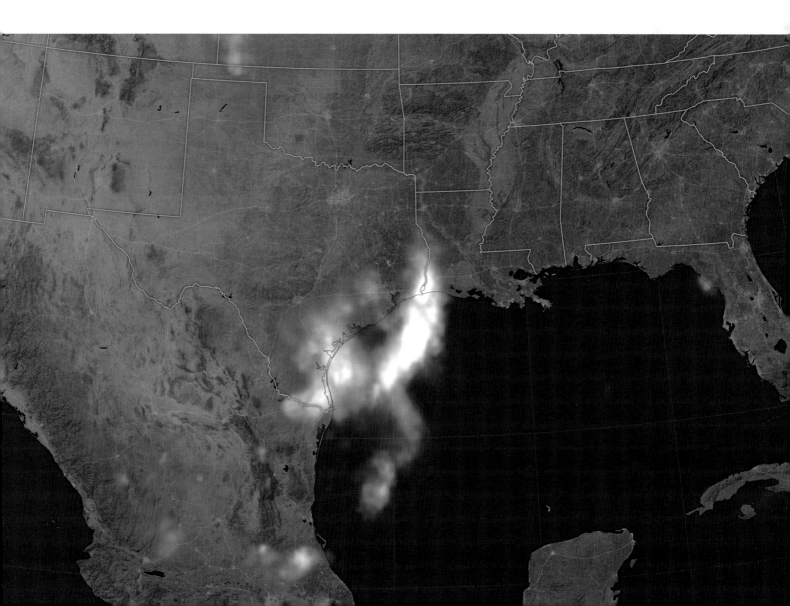

It is perhaps the most viscerally limiting of resources; we can live without a lot, but we cannot live without water. For those reasons, we have gone to great lengths to reduce variability in both water stocks and deliveries, buffering low rainfall periods with reservoirs, building immense canals to transport water, and generating a complex web of infrastructure to balance out fluxes of water to thirsty people and crops throughout arid regions.

FRESHWATER RESOURCES: CALIFORNIA AS A CASE STUDY

A particularly interesting example of water management is the state of California. The world's 5th-largest economy, California is a powerhouse of agricultural and service industries. Industrial farming in California requires irrigation; family and other types of farming require people. All require water. Central California, the heart of the state's agricultural empire and tech industry, is drained by two mighty rivers, the San Joaquin and the Sacramento. The San Joaquin is the longest river in the southern half of California, the Sacramento is the longest in the state. Both meet in the Sacramento-San Joaquin River Delta, one of the most biologically rich areas on the planet, a series of freshwater marshes that formed as a result of water piling up behind a narrow outlet to the sea, plugged with sediment coming down from the mountains. People have inhabited the region for at least 10,000 years. Early settlers noted the bounty as well, and the area was rapidly developed for agriculture and later residential and urban water use. Initially, the inherent variability in fresh water was not a large-scale problem; the wide drainage area of the watershed (all of central California!) meant a drought in one area was often offset by heavy rains or snows in other areas. But as the population increased, periods of low water began to squeeze. Folks looked for water where they could find it. When prospectors flocked to California for fame and fortune in 1849, Tulare Lake was the largest lake west of the Mississippi, surrounded by sedges and grasses 10 ft. (3 m) tall. Tons of aquatic life swam among the reeds, elk foraged along the edges. The local Indigenous communities developed a unique boat with holes on the bottom for spearing fish. But Tulare was primarily seen as a resource to extract and utilize, a danger to be contained, or both. Tulare Lake is gone now, dammed to stabilize floods, partitioned for agriculture, and drained for irrigation and thirsty cities. Most people do not even realize what was lost.

Wresting control of the Tulare area river and lake system was one of the spiritual beginnings of the California State Water Project. A need to support the

Prior to mass agriculture, the marshlands of central California stretched hundreds of miles north and south of the San Joaquin and Sacramento Rivers' confluence, one of the most fertile natural regions in the world (in this map from 1873, wetlands and marshes are denoted as "overflow lands" in gray hatching; early canals are thin black lines). In the south, Tulare Lake presided over the Central Valley and supported vast reed beds and wetlands. Together, more than 300,000 Indigenous people are estimated to have thrived in California prior to European settlement. Today the natural ecosystem is largely gone, diverted and drained.

MAP
OF THE
SAN JOAQUIN, SACRAMENTO
AND
TULARE VALLEYS
STATE OF CALIFORNIA
prepared under the direction of the
BOARD OF COMMISSIONERS ON IRRIGATION
appointed under the Act of Congress approved
March 3rd 1873.
showing the country that may be irrigated and a
PROVISIONAL SYSTEM OF IRRIGATION
Compiled from the Maps of the
GEOLOGICAL SURVEY OF CALIFORNIA
and from
Special Surveys and Examinations
Scale 1 inch to 12 miles
1873.

Published by authority of the Hon. SECRETARY of WAR
in the Office of the CHIEF of ENGINEERS U.S.Army.

burgeoning population in southern California drove people to seek water wherever it could be found. After initial plans to tap rivers in the far North and divert them south via the Sacramento River faltered, dams began to rise on rivers throughout central California, dams that would store water for periods of low rainfall in more arid regions—releasing the effects of natural variability in snow, rain, and atmospheric river events in a more stable and controlled fashion. The water was (and continues to be) diverted into an immense network of artificial rivers, aqueducts, canals, and tunnels that moved water in, around, and under mountains and hills. The natural marshes and sloughs, ever shifting their courses, were channelized to provide regularity. Humans celebrated their ingenuity. New plots of land, marshland drained for easier farming and surrounded by these new canals, took on names reminiscent of very different places: Staten Island, Coney Island, optimistic monikers like the New Hope Tract, or Prospect and Rough and Ready Islands. The land was engineered into one of the most efficient farmlands in the world, hot and sunny weather with ample water, regularly supplied. (When the snows and rains failed, agricultural communities looked to groundwater for supplemental moisture. Underground pumping of fresh water was so intense that large swaths of farmland have actually dropped around 28 ft. or 6.5 m, as the dried subsoil compacted beneath crops.) The net result today is a web of canals and wells built over the last century, each one a little capillary or large artery for the lifeblood of a region.

THE NET RESULT TODAY IS A WEB OF CANALS AND WELLS BUILT OVER THE LAST CENTURY, EACH ONE A LITTLE CAPILLARY OR LARGE ARTERY FOR THE LIFEBLOOD OF A REGION.

Meanwhile, freshwater organisms struggle to adapt to their new host. Salmon, the living counterflow in the region, now migrate upstream through this maze of downflowing waterways. Four distinct king salmon runs still struggle through these waters, including a unique winter-run species that passes under the Golden Gate Bridge in midwinter to spawn in summer, the warmest part of the year. Historically, this was possible because the cold snowmelt of the Sierras would pulse down at exactly that time, keeping water temperatures tolerable for the eggs and smolt. But the vast infrastructure of canals, ponds, pools, and reservoirs—built to contain the flush of melt for use in the summer months—means standing water, waiting to be used, warming in the sun. After the dam and diversion construction really got underway in the 1960s, the salmon run fell from 120,000 spawning fish to under 200. Prevented from reaching the upper elevations of the Sacramento River, the salmon stacked up behind the new dam. Rather than laying their eggs in cool, spring-fed

By 1932, central California's delta was divided into tracts and farms (especially in San Joaquin County, seen on the right side of this map), distilled into manageable chunks of prime farmland.

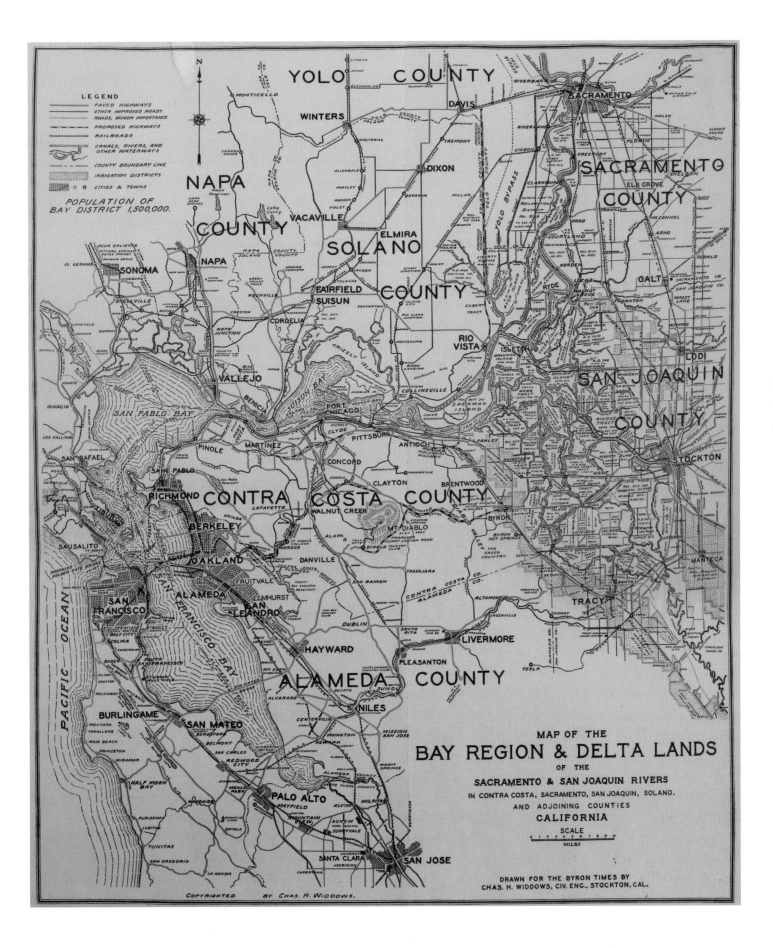

MAP OF THE
BAY REGION & DELTA LANDS
OF THE
SACRAMENTO & SAN JOAQUIN RIVERS
IN CONTRA COSTA, SACRAMENTO, SAN JOAQUIN, SOLANO,
AND ADJOINING COUNTIES
CALIFORNIA
SCALE
MILES

DRAWN FOR THE BYRON TIMES BY
CHAS. H. WIDDOWS, CIV. ENG., STOCKTON, CAL.

COPYRIGHTED BY CHAS. H. WIDDOWS.

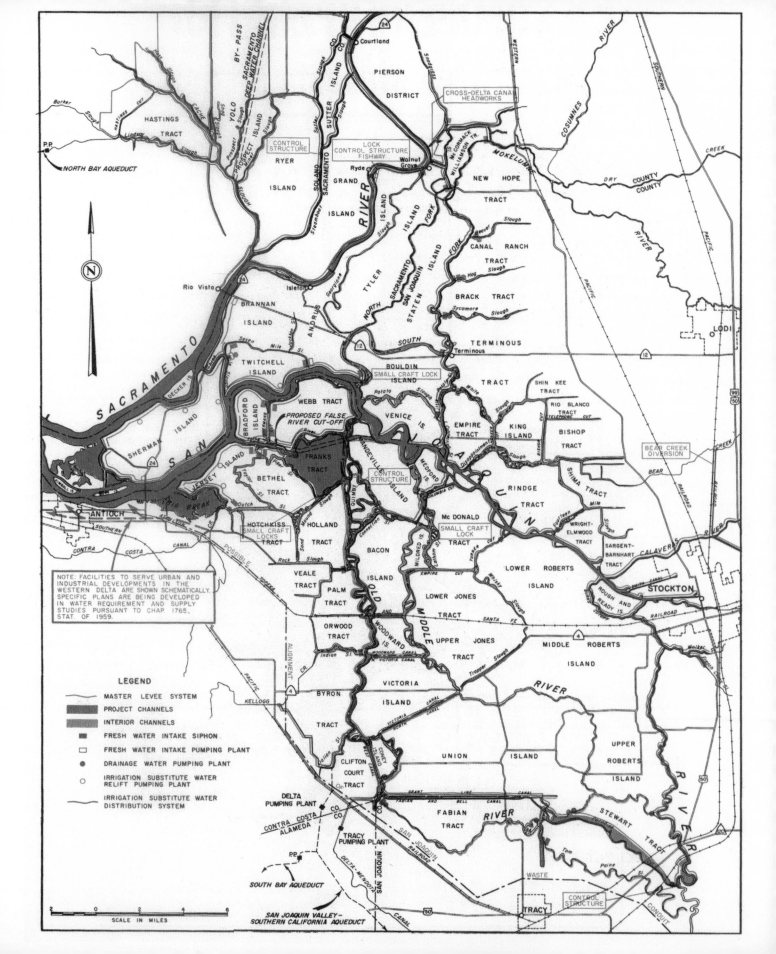

NOTE: FACILITIES TO SERVE URBAN AND INDUSTRIAL DEVELOPMENTS IN THE WESTERN DELTA ARE SHOWN SCHEMATICALLY. SPECIFIC PLANS ARE BEING DEVELOPED IN WATER REQUIREMENT AND SUPPLY STUDIES PURSUANT TO CHAP 1765, STAT. OF 1959.

LEGEND

	MASTER LEVEE SYSTEM
	PROJECT CHANNELS
	INTERIOR CHANNELS
	FRESH WATER INTAKE SIPHON
	FRESH WATER INTAKE PUMPING PLANT
	DRAINAGE WATER PUMPING PLANT
	IRRIGATION SUBSTITUTE WATER RELIFT PUMPING PLANT
	IRRIGATION SUBSTITUTE WATER DISTRIBUTION SYSTEM

SCALE IN MILES

2 0 2 4 6

LEFT
Early diversions and proposed structures in the San Joaquin delta, charted in this 1960 map, were initial components of the California State Water Project—now one of the largest water management efforts in the world, complete with siphons and pumps, locks and dams. All were means to manage the most precious resource in the region: limited water.

FOLLOWING
Lakes also integrate their surroundings. By around 1694, the scope of the Great Lakes was being realized by European cartographers. This early map, by Coronelli, was the most accurate map of the region. Coronelli was fascinated with communicating large regions accurately; he had previously constructed globes for Louis IV of France that weighed 2 tons and were approximately 12 ft. (3.5 m) in diameter. Mapping the Great Lakes to this level of accuracy, in particular the rivers and waterways so important to navigation, was an accomplishment unsurpassed for over 100 years.

mountain streams, they had to stop at the base of the concrete blockage, their redds potentially exposed to the heat. In droughty years, water can run low enough in the reservoirs that there is no longer a cool pool at the bottom and the water is hot all the way through. This can spell death for young salmon.

The fundamental conflict is that the natural cycle of water and fish does not jibe with the need for consistent water to domestic faucets or irrigated yards and crops, which need water late into the dry season. Lawsuits over fish rights to water versus human rights to irrigation continue, vitriolic and mean in the courts. "Whisky is for drinking, water is for fighting over," proclaimed Mark Twain; that fight extends from people to states to the fish in the streams.

But Mother Nature is nothing if not flexible, even if our demands on her are not. Recent evidence from a variety of studies shows that our mapping of critical salmon habitat was overly rigid (or the fish have adapted or evolved), and the winter-run fish are starting to branch out into other streams. While they still spawn in their now-constrained Sacramento River, the hatched fry are mobile, and appear to actively seek out better habitat. Just as tiny salmon do along the Yukon River in Alaska and the Fraser River in Canada, these central California juveniles wander—spreading the population around, lowering the risk that any one unfortunate event, such as a landslide or low-flow year, will wipe out the entire fry population. It may even mean more habitat for spawning one day, if restoration efforts are successful. Flexibility is key to resilience, something the salmon have mastered.

A WATERY SUMMARY

The ocean has been a bastion of stability, buffering us from the worst of the warming for decades. Cool ocean breezes make warmer climates tolerable, and ward off the worst of the cold. Wet and warm marine winds support agriculture in latitudes that would otherwise be too cold, too dry, or both, on all continents. The ocean also provides what has often seemed an inexhaustible bounty of food and a natural highway for seagoing trade. The ocean is always there: the Gulf Stream helping to warm the North Atlantic and Europe, the gentle sea breezes cooling southern California. We thrive on predictability. So humanity has built its cities on the edges of oceans: over a third of all humans live within 60 mi. (100 km) of the ocean. Alas, this constant presence and quiet stability has blinded us to the fact that the oceans are changing, slowly but surely. And the dependable nature of that historic stability, the high thermal inertia which has kept the oceans such a steady presence, means that once

270   275   280

50

Tinthonha
Popoli

Issatis Popoli,
che sono' 24 Villagi

Lago di
Buade, ò
di Issati,
Scoperto l'Anno
1680 dal Conte
di Frontenac

Ouia de'
Battons

Kilistinons P.

Assinipoüalaes P.

F. Colbert, ò Mechisipi

F. di S. Francesco, ò di Sioux, e di Nodoussioux

Nadouessi P.

I. Miniong

L. TRACY, ò S

I. de' Detour
I. S. Michele

F. del Profetta

Villa delli
Siou

F. delli Mosoüittns

L. della
Prouiden
za

R. S. Iacques

R. de
Montreal

Portage
Kioüehounaaü

Ance da
Kaonan

L. O

Passo, ò Caduta di
S. Antonio di
Padoua

F. della Madalena

Le Tombeau, ò
il Sepolcro

Villa di
Siou

45

Nantounagans P.

L. des Pleurs, ò
delle Lacrime

R. Ouatebanga

Outaouami

Katon

Portage

Outaga
mi

E   DELLA   L. O

R. Noire, ò F. Negro

Portage

L. S.
Frances

Maskoutens

F. Colbert, ò Mechisipi

R. a Chekagou

R. Ouisconsing

Li 17 Giugno 1673 il P.
Marquette, e Iolliet furono
li primi degli Europei, ch'
entrorno nel F. Mechissipi
per il Fiume Ouisconsing,
essendo partito da Mas-
koutens li 10 Giugno, con
la guida de' due Miamis

F. Pelasoni

F. de' Chekogou

F. Castorbua

F. Keatiki

P   A   R   T   E

Occidente

40

270   275   280

NS DES TERRES

LA LOVISIANA,
Parte Settentrionalle,
Scoperta sotto la Protettione
di LUIGI XIV, Rè di Francia, etc.
Descritta, e Dedicata
Dal P. Cosmografo Coronelli,
All'Illustriss; et Eccellentiss S. Zaccaria Bernardi,
fù dell' Ecc S. Francesco.

Figuitigouche

Le Pix

Massinaigan o Escriture

Teste de Loutre Michipi coton

I. Montreal

F. Botchitawen

Anse à le Perche

ERIOR

I. au Parisien

I. Montreal

Pau Pin

Montagne del Nort

I di Amicoüe

L'Oriente

Sault, o Caduta di Matanon

Sault au Lievre o Les Galats

Sault aux Allumettes

I. Bargogna

Sault des Calumet

Sault des Chati

Nipissiniens Les Missiraghe

I S I A N A

LOVISI

I. Sgnacio

Nipissiniens Les Missiraghe

L'Oronto

F. degli Oultaouacs, o degli Huroni

I. DEGLI

HURONI, detto Algonkins- Mikigang

Toïonagon Canatha Kiagon

Trontenac

F. degli Iroquesi

Nahiha

Onontagueronons

FRONTENA

MICHIGAMI

Cde l'Ours

R. de la Monistaque

B. de' Sukinand

Atiraghenrega

L. di Triketi, ò d'Acque Salse

Saut de Niagara

Innantouan Kakouagga

Tronttuaeronons

Goyegouenronons

L DIGLI ILLINOI, ò

R. Sakinand

LAGO ERIE

Il Lago Erie é altrimente Chiamato
Teïoch-Rontiong, ò Conty, ò du Chat

Forte Miamis

Villadi Miamis

Portage

Miglia d'Italia

60   120   180

Leghe di Francia

37   75

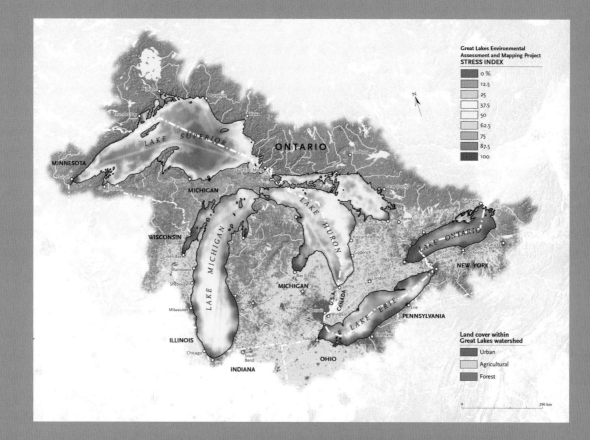

Great Lakes Environmental
Assessment and Mapping Project
STRESS INDEX

	0 %
	12.5
	25
	37.5
	50
	62.5
	75
	87.5
	100

Land cover within
Great Lakes watershed

Urban
Agricultural
Forest

Stress on the Great Lakes

The Great Lakes are the largest freshwater lake system on the planet (by surface area) and 2nd largest by volume (to the monstrous gouge that is Lake
Baikal in Russia). They are rimmed by both urban and wilderness areas, but
their prime location near the heart of the North American continent and the
resultant shipping and industrial pressures meant the human impacts won
out. But some lakes are worse than others. Redder areas in the lakes mean
more stress; blue areas less. Lake Erie, intensely industrialized in the early
20th century, has struggled; Lake Superior, with its still-wild shores in western Ontario, Minnesota, and Wisconsin, has held on so far. Measuring that
impact is tricky; here, stress is a composite of invasive species, industrialization, shipping pressures, commercialization, and other factors. Standing on
the shores of Lake Michigan, one might think these bodies of waters are too
big to be impacted by humans, but the larger view reveals that the fringes of
these freshwater giants are also part of the human-dominated world.

the oceans start to change, that change is nearly impossible to stop. We are already locked into decades of warming because the oceans have warmed appreciably; even if all emissions were miraculously eliminated, the oceans would continue to rise. They would also continue to heat the atmosphere, warming the air. The very nature of oceanic variability—slow and muted, but heavy with each swing—means that climate change impacts on our oceans carry that much more weight.

In contrast, rivers, lakes, and freshwater systems are persistently variable. Floods ravage cities; droughts kill cities. We are drawn to the resource that can kill us through variance, and so we control that variance with man-made contraptions: levees for flood control when there is too much water, canals to carry water when there is too little. Once again, the natural world with all its pulses is mismatched to the human experience and human needs, or at least the needs of the built environment we have created. And that control is fated to fail, either through natural events that exceed our capacity for control, political conflicts that happen when human geographies don't correspond with natural systems, or the ongoing push of climate change.

"Nothing is softer or more flexible than water, yet nothing can resist it," wrote Lao Tzu in the centuries before Christ. As the waters go, so goes the planet.

"NOTHING IS SOFTER OR MORE FLEXIBLE THAN WATER, YET NOTHING CAN RESIST IT."
—LAO TZU

LAND

INFINITE VARIETY

The defining attributes of the terrestrial world are variety, variety, and more variety, also known as heterogeneity. If the atmosphere is well mixed, and the oceans of the world dominated by their ponderous inertia, the land is fickle, fast changing, and incredibly variable. In the Himalayas, 10 mi. (16 km) can take you from the top of the Dhaulagiri massif—the 7th-highest peak in the world at 26,795 ft. (8167 m) and well into the "Death Zone," where low oxygen levels can kill—to the Gandaki River, at 8270 ft. (2520 m), lower than many mountain vacation towns. Villages along this river enjoy a warm, temperate environment with mild winters. Another 10 mi. (16 km) on the same line takes you back up to the peak of Annapurna, 26,545 ft. (8091 m), the 10th-highest mountain in the world. This V-shaped trek travels from killer cold to warm and pleasant back to killer cold in only about 20 miles (as the raven flies)—a distance most people could walk in a day (ignoring the vertical component, of course!). The terrestrial environment is amazingly variable, changing rapidly both in time and space.

The composition of a landscape, in broad terms, comes down to the two major climate compositions we have previously explored: temperature and water (generally precipitation, but also drainage). We will tiptoe through this complex heterogeneity by biome, or major ecological community type. We will tiptoe lightly, for the variety of landscapes is far too rich to be summed up in a few pages—but we'll do our best. The science and the mapping are both fascinating.

WETLANDS: THE "WASTED LANDS"

The first step out of the water onto land is a squishy one, for there is no clear demarcation between where water stops and land begins. Rather, we have a gradient from always wet to always dry. And occupying the moister end of that spectrum are wetlands—lands that are saturated for at least part of the year, including the growing season. Wetlands vary from consistently flooded marshes and river deltas to infrequently covered mountain muskegs (mossy wetlands). The thing that defines them is an abundance of water. My backyard as a child was a half acre of solid wooded wetlands, flooded in all but late summer, and never truly dry. As a child, climbing through the thick willows and alders growing out of soil hummocks, I had the perfect playground—the trees were too spindly for adult weight, barely making a living in the water-world they grew from. Kids, on the other hand, could climb through, just barely above the water in an arboreal jungle gym. Bugs were rampant, but so were frogs (which would keep houseguests awake all night), blue herons, and a revolving flock of migrating waterbirds. The land was a sponge, soaking up water and slowly releasing it over time—and in turn, protecting downstream houses from being swamped by the wet weather of Pacific Northwest winters. By serving up a little bit of all habitats at various times, from water, soil, and trees to open, grassy areas with lots of sun, wetlands in their entirety are something unique themselves, an invaluable and highly productive piece of our world.

Wetlands can be broken down into many categories, each with their own idiosyncrasies, charms, and attributes: bogs and muskegs (mossy and cold), fens (fed by

THE FIRST STEP OUT OF THE WATER ONTO LAND IS A SQUISHY ONE.

groundwater), swamps (dominated by woody plants), marshes (dominated by grasses and soft-stemmed plants), tidal (influenced by saltwater), and more. But they all share an outsized role in the story of life. The abundance of languid, lazy water and consistent inputs makes these systems as productive as tropical rainforests, incredibly biodiverse, and fragile. They slow things down, soaking up floodwaters like sponges. That slow, still water can then percolate through the soil, giving a myriad of bacteria time to process hard-to-digest materials such as our modern pollutants. What exits is cleaner and more regulated then the wild flows that come in.

The diversity of wetlands grows from the fact that productivity is very high (plentiful water, light, and nutrients from incoming water), decomposition is low (oxygen-poor conditions), and those factors are highly heterogeneous, meaning they vary quickly across space; a bump of only an inch or a couple of centimeters can mean the difference between completely submerged and occasionally dry—and that means diversity in the plants and animals that specialize in different conditions. The high production of biomass feeds a variety of insects, fungal species, and both lower and higher plants. Cathedral-like cypresses can live thousands of years, presiding silently over the lazy spinnings of their swamps. In the lower 48 of the United States, some two-thirds of all waterfowl reproduce in the wetlands of the Midwest. In the global North and South, muskeg ("turbal" in South America) regions are a spongy mass of *Sphagnum*, so consistent that you can dig down 10 ft. (3 m) to find leaves 1000 years old that look like they grew last year. Tropical forested wetlands store more carbon from our atmosphere than almost any other place on the planet. They are as diverse, and as valuable, as you can imagine.

Unfortunately, the recent history of wetlands and humans is not a tale of conviviality and has turned adversarial in many cases. Yet this was not always the case; wetlands have provided resources for locals for thousands of years. Labrador tea (actually three closely related *Rhododendron* species) has nourished Indigenous peoples in the far North; the Peul (or Fula) peoples in Senegal have seasonally farmed riverine wetlands in the midst of dryland Africa, allowing for a consistent supply of food despite the intrinsic variability of their environment. Coho salmon in Alaska dart around the connected pools of forested wetlands as they make their way to the sea. Upon their return, they support vibrant Native populations, as they have for millennia.

But with the rise of modern societies and their attendant industrialized agriculture, population density, and building patterns, the variability of wetlands—sometimes flooded, sometimes not—became a detriment. The high productivity and

FOLLOWING
Differences in warmth and water produce vast amounts of variety on land, physically reflecting the patterns of temperature and precipitation produced by atmospheric circulation at the global scale. From frequently submerged wetlands to windy mountaintops, the combination of latitude, altitude, and topography drives the endlessly fascinating geography of life.

LIFE OF THE LAND: VARIETY ON VARIETY

The myriad of environments on land are primarily caused by different amounts of just two things: water and heat. The amount and variability of both matters, from the persistently damp wetlands to the cold winters and short growing seasons of the tundra.

BIOMES

Mountains

Tundra

Grasslands

Tropical forests   Temperate forests   Boreal forests

Deserts

Wetlands

Kali Gandaki Gorge

KALI GANDAKI GORGE

The Kali Gandaki Gorge, a wonder of the world, is the deepest in the world. The Gandaki River cuts between two 26,000-ft. (8000-m) peaks separated by only a few miles on its way from the Mustang border region of Nepal and Tibet (China) to the Indian Ocean. At the river bottom, warm summers bathe local villages and farms; at the top, low oxygen, ice, and snow mark the famous "death zone" of the Asian mountains.

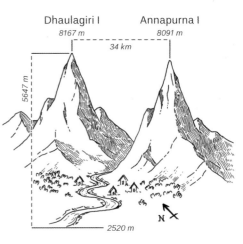

Dhaulagiri I
8167 m

Annapurna I
8091 m

34 km

5647 m

2520 m

N

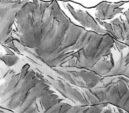

Annapurna I

Annapurna South

Dhaulagiri I

complex nutrient processing in wetlands creates amazing soils for farming, and the lure of control became irresistible. Wetlands were drained and planted; it is remarkably easy to destroy land dependent on the slow percolation of water—just remove the water. A simple ditch is enough to lower the water table and rapidly sluice the essence of these wetlands but leave rich soil for the taking. The previously mentioned Tulare region, formerly rich wetlands and lakes, is now one of the most productive agricultural landscapes in the world. The soil pays dividends, but at a cost.

In some cases, wetlands were drained for human health benefits as well. Malaria has long been tied to wetlands, taking its name from the *mala aria* ("bad air" in Italian) associated with the swamps of northern Italy, but tracing its evolutionary history further down the human chain—from Rome to Egypt to the evolution of *Homo sapiens*. As recently as 1912, malaria was common in Washington DC, spread by mosquitoes that bred by the millions in freshwater swamps. In 1942, the Office of Malaria Control in War Areas was established; in 1947, it was repurposed toward the elimination of malaria in the United States. Through a warlike campaign, pesticides (DDT at the forefront among them), house control, and coordinated draining of wetlands to remove breeding habitats took on the disease directly, and by 1952 the war was over. Between 90 and 94 percent of coastal wetlands from Maine to Virginia had been ditched, not an unusual number for the effort. And it worked: malaria has effectively been removed as a pathogen in the United States, with only around 2000 cases a year, mostly attributed to travelers from other counties.

Drainage of wetlands for agriculture was common practice in Europe also; from 1895 to 1995, Europe drained and filled two-thirds of what was left. Globally, estimates range from 54 to 87 percent in terms of how much natural wetland area has been destroyed since the year 1700. The loss continues apace; the US has slowed to a loss of approximately 0.5 percent of area per year, but Europe, Africa, and Oceania are losing natural wetlands at about double that pace (one percent per year), Asia at triple the rate (1.5 percent per year), and the neotropics at a staggering 2 percent per year. And while humans also create wetlands such as rice paddies and fringes of reservoirs, the amount gained is far outweighed by the losses.

Unconsciously, we miss the benefits of natural wetlands, the water purification, the habitat, the "wildest and richest gardens that we have," according to Thoreau. "Hope and the future for me are not in lawns and cultivated fields, not in towns and cities, but in the impervious and quaking swamps." Thoreau was referring to the essence of life, not sterile but teeming with squirming and straining organisms. Every individual wants

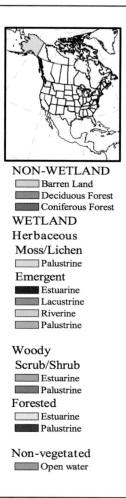

NON-WETLAND
- Barren Land
- Deciduous Forest
- Coniferous Forest

WETLAND

Herbaceous

Moss/Lichen
- Palustrine

Emergent
- Estuarine
- Lacustrine
- Riverine
- Palustrine

Woody

Scrub/Shrub
- Estuarine
- Palustrine

Forested
- Estuarine
- Palustrine

Non-vegetated
- Open water

DRAINAGE OF WETLANDS FOR AGRICULTURE WAS COMMON PRACTICE.

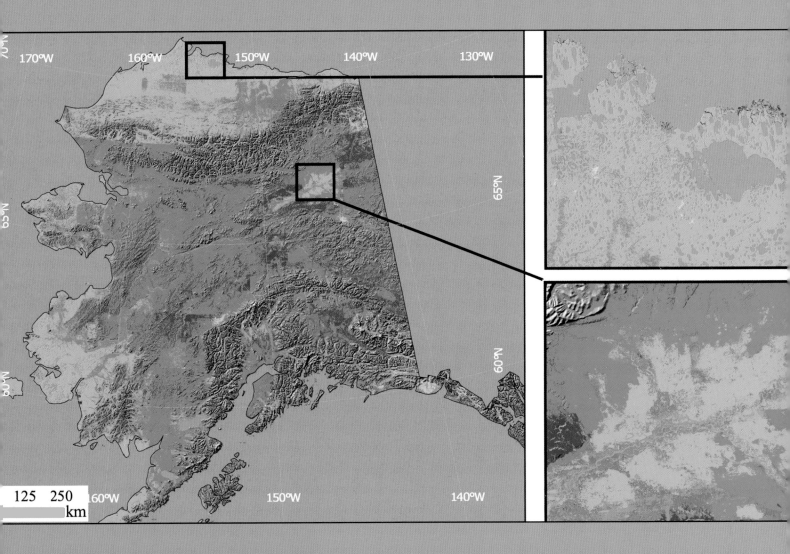

Alaskan Wetlands

Alaska is about 36 percent wetlands, a remote and beautiful region of muskegs, bogs, and marshes. Estimates vary by mapping method; this map by Daniel Clewley and colleagues from 2015 charted 0.23 million sq. mi (0.59 million sq km) across the state. The variety of wetland types (palustrine, where the water comes from groundwater; estuarine, from ocean water; lacustrine, from lakes; and riverine, along rivers) are scattered throughout.

EBBING WETLANDS

Gulf of Mexico

Lake

THE E

BIG CYPRESS

NATIONAL PRESERVE

Everglades City

TEN THOUSAND ISLANDS

5.2 feet of
sea level rise

2.5 feet of
sea level rise

E V E R G L A D E S

Present-day
average sea level

Shark Riv

Key McLaughlin

SEAS ON THE RISE

Scientists say sea levels could
rise 2.5 (••••) to 5.2 (‐‐‐) feet
by century's end, causing
much of the Everglades to
be inundated by salt water.

This map uses false-color
infrared satellite imagery
to highlight land cover that
could be transformed.

Mangroves are important to
coastal integrity and thrive in
areas inundated with seawater.
But as ocean levels increase,
they spread inland, replacing
freshwater wetland plants.

What look like speckles within
the Everglades are small **islands
of trees** rising above the freshwater saw grass wetlands.

Vegetation strongly reflects
infrared energy, absorbing red
and blue wavelengths in the
process of photosynthesis.

Features shaped by **human
activity**—roads, buildings,
even farmland—absorb sunlight,
then release the energy as heat,
leading to a bright appearance.

Water features absorb the sun's
energy, leading to an inky black
appearance in this view.

Broad

N A T I O N A L

Long

Harney

THE

Shark

*Ponce de Leon
Bay*

P A R

Whitewater Bay

Taylor

C A P E   S A B L E

⊞ Flamingo
Visitor Center

F l o r i d a

B a y

Florida's famed wetlands, the Everglades, are pinched between a burgeoning Miami to the east and encroaching saltwater to the west. With sea levels rising, the immense freshwater marsh hangs in the balance. By 2100 most of this unique national treasure could be dramatically altered.

BY MATTHEW W. CHWASTYK

Flooded quarries

Hialeah

Miami Beach

MIAMI

5.2 feet of sea level rise

eet of el rise

Extent of Greater Miami built-up area

Coral Gables

Key Biscayne

Kendall

UP TO THE EDGE
Urban and agricultural areas of Greater Miami press directly against the protected Everglades. Agricultural pollution from as far as 50 miles north harms the wetlands.

Biscayne

Bay

Agricultural area

Elliott Key

Homestead

Turkey Point Nuclear Generating Station

2.5 feet of sea level rise

F. Coe Center and eadquarters

Flooded quarry

Old Rhodes Key

Card Sound

Key Largo

ATLANTIC OCEAN

nt-day ge sea level

Barnes

Sound

0 mi   6
0 km   6

Key Largo

Coastal wetlands—those that remain after development, draining, and removal for agriculture—are also under threat from rising seas, illustrated once again by southern Florida, where the iconic wetlands of the Everglades are beginning to ebb away.

A Restoration Success Story

The value of wetlands is being recognized in developed landscapes around the world, a realization that this natural system is a uniquely important part of a functioning ecosystem. The Tres Rios Wetlands, where the Salt, Gila, and Agua Fria Rivers join in Phoenix, Arizona, is one such place. Formerly a flashy series of rivers that were either dry or flooding, the landscape was drained and leveled for agriculture. Now, after intensive reclamation, invasive species removal, and terraforming to restore the waterflow, the rejuvinated landscape is home to native rushes, sedges, trees, and floating plants. It hosts incredibly diverse bird communities. Plant water use, especially in the summer, is so high in the hot desert environment that it pulls water from open ponds into the vegetated marsh during the day, what researchers call a "biological tide." The plants absorb excess nutrients (themselves pollutants) as well as oils and other synthetic contaminants. This makes the Tres Rios an incredibly efficient pollution-removal system and an invaluable component of the functioning landscape.

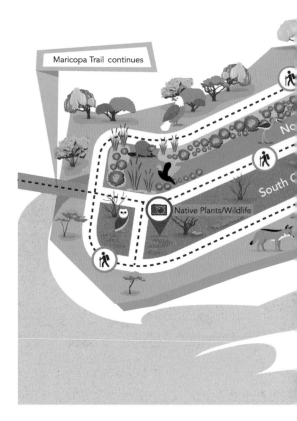

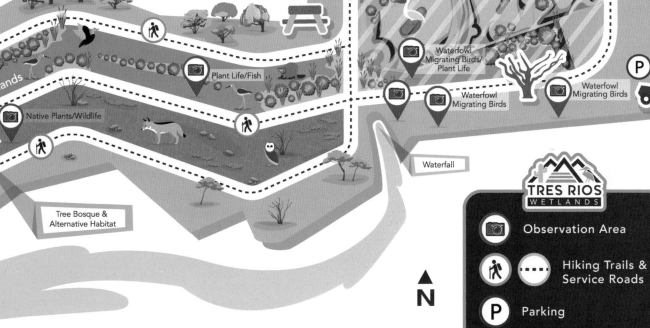

About The Overbanks

The North Overbank Wetlands (OBW) channel is designed to be a perennial segment characterized by emergent marsh, open-water, and riparian forest. Water will combine into a single channel and flow until it encounters one of 8 internal deep zones. This pattern of emergent zones and open-water deep zones repeats itself for the length of the channel.

The South Overbank Wetlands (OBW) channel is designed to be an ephemeral system characterized by mesquite bosques, riparian gallery forest, and riverine terrace vegetation. Water can be diverted to flood irrigate cottonwood/willow riparian areas and mesquite bosque areas through the use of irrigation discharge structures. Excess irrigation water will flow to the center of the channel and be collected into a meandering swale and travel downstream to the facility outlet and/or infiltrate into the soil.

Please visit phoenix.gov/waterservices/tresrios for more information and to request an access permit.

Roeser Road

CLOSED TO THE PUBLIC

Waterfowl Migrating Birds

Waterfowl Migrating Birds/ Plant Life

Waterfowl Migrating Birds

Waterfowl Migrating Birds

Plant Life/Fish

Native Plants/Wildlife

k Wetlands

etlands

Tree Bosque & Alternative Habitat

River

Waterfall

S. 91st Avenue

TRES RIOS WETLANDS

Observation Area

Hiking Trails & Service Roads

Parking

Closed to the public

N

Forests are built on complexity and life takes
advantage, from ephemeral wildflowers such
as trillium, to flowering head-high azaleas, to
towering oaks and pines.

Flowering Dogwood

clean water, flood control, and habitat; we turn away from stagnant ponds laced with weedy overgrowth from pollutants, from fatty sheens over dry hardpack. We lament the coastal nutrient pollution that creates dead zones on our waterfronts, driving away fish and sliming our swimming holes. Wetlands are our partners in avoiding those losses. Wetlands, by virtue of their nature—one foot in the water, the other on land—perform functions that we cannot get elsewhere, the critical nexus of life that cleanses some of our sins of pollution. You have to bend over and look closely; the wonders of wetlands occur on the tiniest of scales, but with global implications—and we continue to reap those benefits, as long as we can keep them intact.

FORESTS: REACHING THE LIMITS OF LIFE

> For me, trees have always been the most penetrating preachers.... Nothing is holier, nothing is more exemplary than a beautiful, strong tree. When a tree is cut down and reveals its naked death-wound to the sun, one can read its whole history in the luminous, inscribed disk of its trunk: in the rings of its years, its scars, all the struggle, all the suffering, all the sickness, all the happiness and prosperity stand truly written, the narrow years and the luxurious years, the attacks withstood, the storms endured. And every young farmboy knows that the hardest and noblest wood has the narrowest rings, that high on the mountains and in continuing danger the most indestructible, the strongest, the ideal trees grow.
> —Hermann Hesse, *Trees: Reflections and Poems*

Forested landscapes capture the imagination. From the dark, Big Bad Wolf–infested hills of Bavaria to the dry, cowboy-infested woodlands of Montana, forests have their own character, feel, and temperament. Humans have a special relationship with forests and trees, perhaps spiritually harkening back to our arboreal past and practically a result of the utility of wood, the animals that thrive in trees, and forests' role in the water cycle. They are messy ecosystems: layers and layers of structure, jumbled, leaning, and stretching both up into the sky and down into the soil. The reason they are so biodiverse, so valuable as habitat, is in large part due to this complexity and height. Trees are not necessarily long lived, not in the big scheme of things; some trees can grow and reproduce in as little as a decade. Others last thousands of years, but that is still relatively short in the context of the world. Plants grow on plants (termed "epiphytes") and have a lot of choices in forests—high in the canopy for maximum light, low on the ground for cool shade. Animals are similar; small mammals may spend their entire lives in the canopy, without once touching the dirt. Others burrow among the roots, finding protection in the woody hollows that twist through the upper soil layers. The magic of forests thus comes from complexity, from variability. Growing in nooks and crannies are plants with medicinal value, food value, spiritual value. But the defining attributes of forests, whether in Siberia or Brazil, are their height and structural complexity, the jumble.

FOREST BIOGEOGRAPHY

Forests occupy a particular spot in climatic space. Temperatures have to be warm enough for trees to grow away from the ground (the sun heats the ground surface more than the air, so staying close to the ground means staying warmer in cold landscapes). As mentioned previously, you can more or less approximate global tree line quite well by the 50°F (10°C) summer isotherm, meaning the line around the world where average summer temperatures are at least 50°F. The landscape must also be wet enough for trees to survive. Many trees have deep taproots and are fairly resilient to low-water conditions, but there is a minimum, which is highly species dependent, and early life is the hardest for a seed. But if they get started, trees can be quite tolerant of variable climates.

Once established, trees dominate. They shade out grasses and other more height-challenged competitors—that is one of the main reasons to grow tall. It is rare to find a nice lawn underneath a thick canopy. But grass isn't entirely helpless, and early on, grasses are typically much stronger competitors for water than tree seedlings. One-on-one, seed versus seed, grasses will carpet the ground and exclude any trees from infiltrating the soil. If the environment is wet enough, though, that advantage goes away—hence the widespread domination of forests at higher latitudes and the true tropics, where water isn't a limiting factor. But if it is dry or cold (in the summer) enough, trees cannot compete for water and lose out to shrubs or scrub vegetation. Think tundra or desert. However, there is an interesting middle ground, too, in which grasses and trees battle for domination. Here either could dominate, and what you see is the product of what ecologists call exogenous factors, things originating outside the system or just random historical chance.

Coastal forests in Mexico are mapped here in three dimensions from a satellite laser altimeter system. The horizontal line (relative elevation) shows the path of the laser; the lower elevational profile shows the slow descent of the land to the ocean, then the flat level of the sea (plus some waves). The trees are the spikes of higher "elevation" values above the land; sharp changes in forest cover are apparent as the lower profile moves through developed areas en route to the water.

Laguna del Mar Muerto

Punta Flor

Pacific Ocean

5 km

N

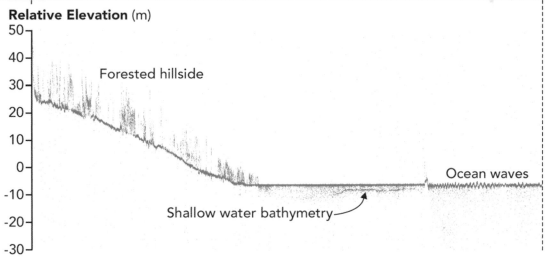

Relative Elevation (m)

Forested hillside

Ocean waves

Shallow water bathymetry

It took more than
three thousand years to
make some of the trees
in these western woods
… Through all the won-
derful, eventful centuries
since Christ's time—and
long before that—God
has cared for these trees,
saved them from drought,
disease, avalanches, and a
thousand straining, level-
ing tempests and floods;
but he cannot save them
from fools.

—John Muir,
Our National Parks

Even as early as this
1873 map of forest
density (measured as
acres of forest per sq.
mi., with darker colors
indicating denser forests),
the deforestation of the
northeastern United States
for agriculture is apparent,
with dense forests limited
to upland New York State
and Maine.

Pl. III.

Pl. IV.

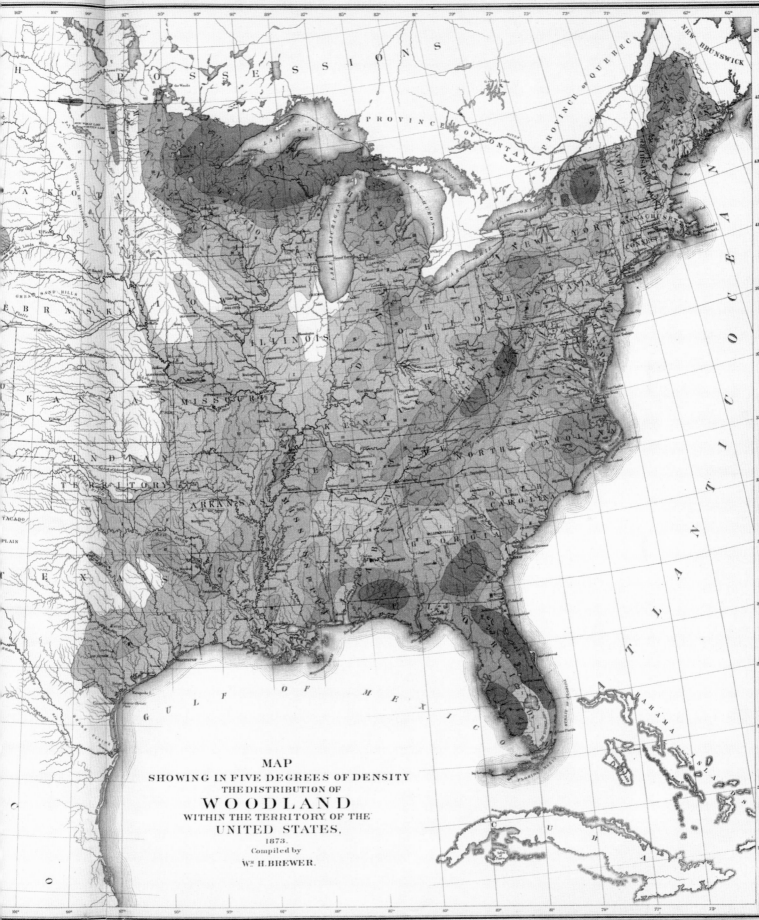

MAP

SHOWING IN FIVE DEGREES OF DENSITY
THE DISTRIBUTION OF

WOODLAND

WITHIN THE TERRITORY OF THE
UNITED STATES,
1873.
Compiled by
Wᵐ H. BREWER.

96
72
48
24 species per
square mile
1

NUMBER OF TREE SPECIES*
per square mile

total species cataloged:
316 species* in the US,
92 in Canada

Forestation Patterns

The forests of North America have a split personality. In the East, there is high diversity and a dominance of broad-leaved, deciduous species. In the West, low-biodiversity conifer forests brave the dry interior and blanket the maritime West Coast. Broad, continental patterns are readily apparent. The floodplain of the Mississippi River, many times as wide as the river itself, guides deciduous trees through the piney fringe along the Gulf Coast—evidence of recent flooding and rich riverine soils, which disfavor conifers. The sudden elevational uptick of the Rocky Mountains means trees can survive in dense forests, unlike the dry, shortgrass steppes just east and below. The folds and "hollers" of the Appalachians (southeast of the Great Lakes) are home to some of the highest tree biodiversity on the continent. Similar but unique; not all forests are created equal, and they are all shifting with climate.

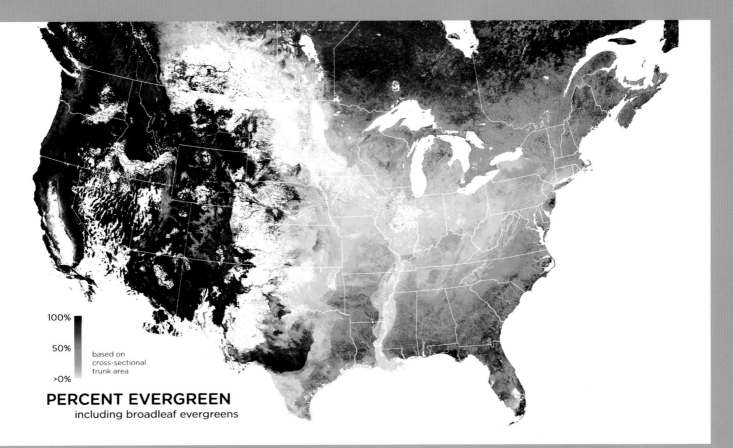

100%

50%

>0%

based on
cross-sectional
trunk area

PERCENT EVERGREEN
including broadleaf evergreens

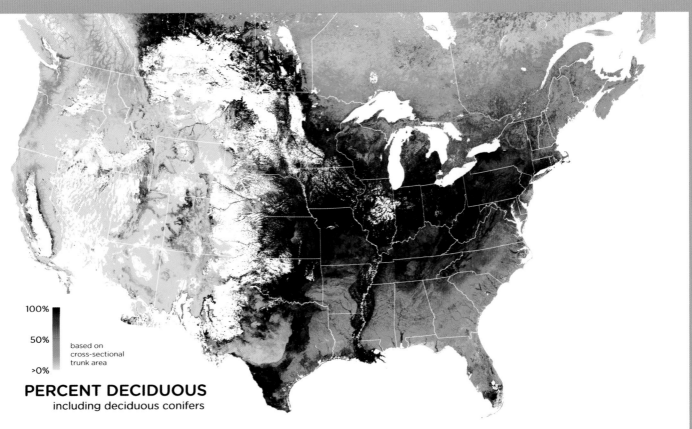

100%

50%

>0%

based on
cross-sectional
trunk area

PERCENT DECIDUOUS
including deciduous conifers

In the case of the African savanna, those factors are wildfires and seed establishment. In that system, wildfires feed on grass, like herds of wildebeests, consuming vast amounts of biomass rapidly. But because grass has a relatively low density, the fires are fast and not particularly destructive. Tree seedlings are killed off, of course, but the grass itself bounces back the next growing season. That means fires can occur every few years with little consequence to the grass, while holding the forest at bay. But if, for whatever reason, there is a gap in the fires for a few decades, you may see a few trees pop above the grass layer. Newly tall, they are less likely to be killed in a low fire and their shade begins to push back the grasses. That reduces the probability of fire a bit (less grass to burn), and more trees can pop up. Eventually, a forest can take over. It's an "alternate stable regime," an entirely different but equally stable ecosystem. Hence, the mosaic of grass and trees that dot the iconic subtropical latitudes of Africa and elsewhere.

There are other considerations—photosynthetic molecular machinery, for example, or wind exposure or soil type, but those are primarily fine-scale factors. The global biogeography of forests can be coarsely described by low-moderate to high water availability and moderate or warmer growing-season temperatures in summer, coupled with the temporal variability of fires. Masters of this pleasantly temperate realm, forests blanket about 30 percent of the planet's land in their thick-covering towers of temporary stability. Perhaps it is no surprise that humans and trees have such a deep bond, seemingly more than just a utilitarian relationship of wood, fiber, and food. Our connection to the woods is a shared link with climate; we like the same things. Their environment is our environment. Their world is our world.

THE DECLINE AND FALL OF THE FOREST EMPIRE

To describe how forests are changing, we need to talk not only about climate but also about human activity and deliberate manipulation. Of all the ecosystems in this book (except the urban system), forests are managed perhaps the most intensively. The origins of forest management depend on the scale of investigation: many argue that there is evidence of intentional promotion of palms in Amazonia in prehistorical times (though the widespread impact is debatable). Tribes in the Pacific Northwest planted crabapple orchards along riversides as gallery forests. And much of western Europe was cleared for agriculture before the fall of the Roman Empire. Fire was an extremely common tool for widespread forest removal and a form of wild agriculture (for example, cultivation of the camas in Washington State). There have been notable achievements in conservation—from the rudimentary Forest Laws of William the Conqueror (not equitable from a human standpoint, but a form of early forest protection) to recent broadscale reserves in Patagonia. Trees are an integral concern of the human enterprise.

Overall, though, the net direction of forest cover change must go either up or down, and it has been downward. Native forest loss has been the story for millennia. Over the last 1000 years, forest area has been reduced by half. The majority of that (about 75 percent) has been in the last 200 years, and in the last 25 years we have lost a little over 3 percent of the total remaining forest area, from 10.1 billion acres (4.1 billion hectares) to just under 10 billion acres (4 billion hectares) remaining. Note that those numbers do not take into account domesticated forests (such as forests maintained for palm production or timber); they only speak to complete loss, such as forest to farmland conversions. The majority of recent forest loss has occurred in the tropics (Brazil, Indonesia, and Myanmar primarily). About 75 percent of this is due to human activity, roughly split between forestry, commodity production, and agricultural clearing. Higher latitudes have actually recorded forest gains over the past few decades; China leads with a net rate of 0.8 percent growth (3.7 million acres or 1.5 million hectares gained) between 2010 and 2015. In the United States, net forest loss stopped relatively early, around 1910, with the establishment of the National Forest System. While individual regions still saw massive destruction of forest ecosystems, and their replacement with industrial, so-called forest agricultural systems, the mass movement of rural populations to urban centers resulted in forest regeneration across wide swaths of formerly cleared land, especially in the Northeast. So while US forests have continued to be industrialized, trees (if not true forests) have gained in area over recent years.

> OVER THE LAST 1000 YEARS, FOREST AREA HAS BEEN REDUCED BY HALF . . . ABOUT 75 PERCENT HAS BEEN IN THE LAST 200 YEARS.

Climate change threatens, however. Being masters of a particular climate space means problems if that climate shifts. From 2011 to 2019, 150 million trees died in California alone due to drought (part of that mortality results from the forests being overly thick due to decades of fire suppression, which led to a higher demand for water that didn't arrive; part of it is increasing temperatures causing more water stress). Insect outbreaks, windstorms, and fires are increasing in frequency and severity; while many forests specialize in fires and fire recovery, no forest can persist with severe fires every decade. Estimates are a 40 percent reduction, or at least simplification, of forests, with a rise in global temperatures of 5.4°F (3°C), which is highly likely in the coming decades. The actual change may lag temperatures a bit—trees are tough creatures—but they are not invincible. And when change happens, it can happen fast.

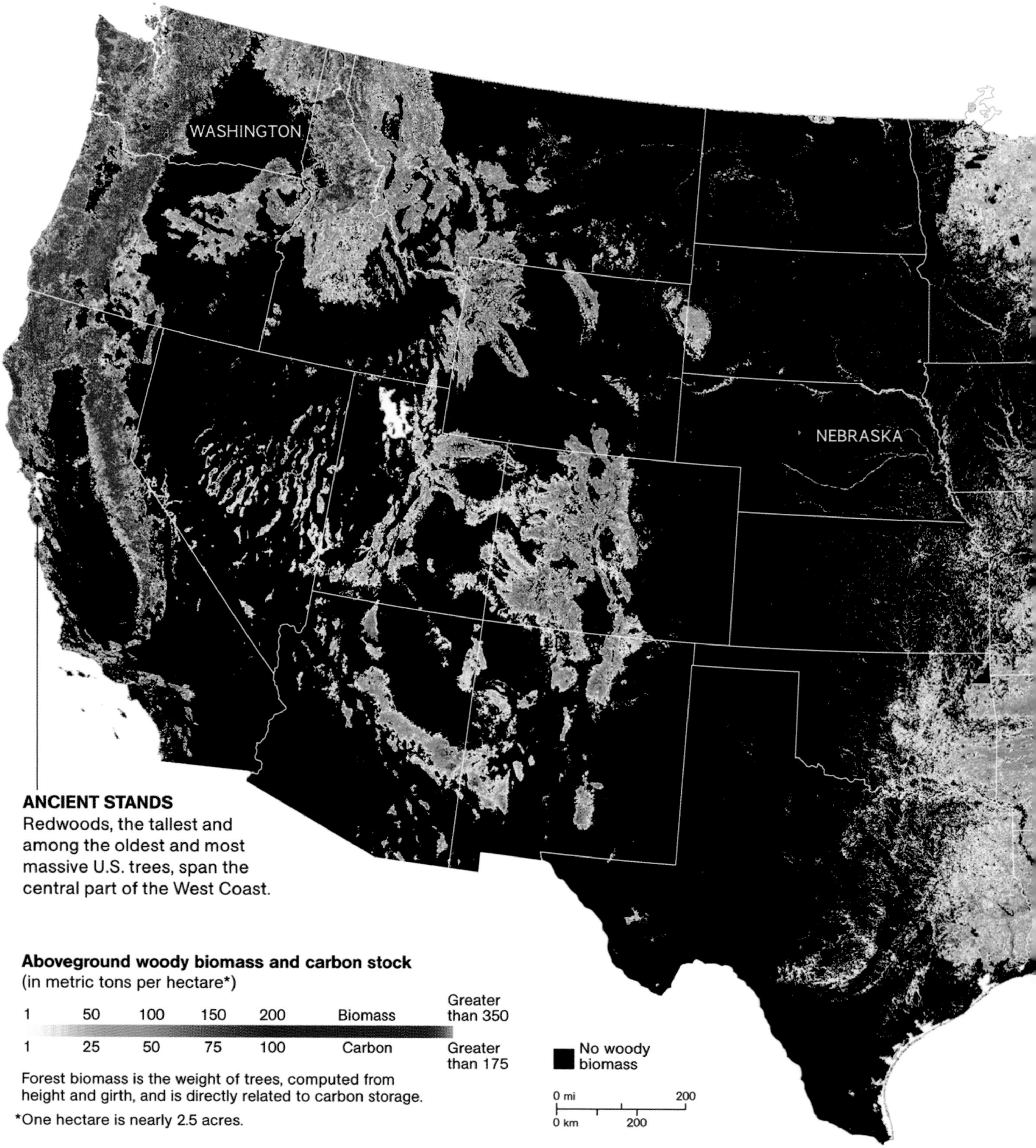

WASHINGTON

NEBRASKA

ANCIENT STANDS
Redwoods, the tallest and
among the oldest and most
massive U.S. trees, span the
central part of the West Coast.

Aboveground woody biomass and carbon stock
(in metric tons per hectare*)

						Greater than 350
1	50	100	150	200	Biomass	
1	25	50	75	100	Carbon	Greater than 175

Forest biomass is the weight of trees, computed from
height and girth, and is directly related to carbon storage.

*One hectare is nearly 2.5 acres.

■ No woody
biomass

0 mi   200

0 km   200

MAINE

FOREST BANDS
Development
proceeded in
stripes along
the Appalachian
range. Woods
remain on slopes
too steep for
building.

TREE FARMS
In the Southeast
fast-growing
trees are planted
as crops. Some
softwoods,
like pine, are
harvested every
20 years.

The legacy of historical
harvest, farmland
abandonment (resulting
in forest recovery in the
Northeast), and the new
prioritization of forest
carbon as an important,
ecosystem-based service
at the national scale are
readily apparent in more
recent assessments of
forest distributions, such
as this one.

RICH FLOODPLAINS
he Mississippi River fertilizes
wide corridor where forests
ive way to corn, cotton, and
oybean fields.

Mississippi

MAP: JEROME N. COOKSON, NGM STAFF; GREG FISKE, WOODS HOLE RESEARCH
CENTER; THEODORE A. SICKLEY. SOURCES: NASA; U.S. FOREST SERVICE; USGS

Attack of the Mountain Pine Beetles

The mass outbreak of mountain pine beetle in the early 2000s, which can kill multiple species of pines in North America, was unprecedented in scope and extent, spanning from northern British Columbia to the desert mountains of New Mexico. The species is not invasive; it is a native beetle. The beetles kill healthy trees only when the insect's populations get extremely high. A combination of extensive drought, warming winters, hotter summers, and extensive, homogenous forests (caused by humans and wildfires) led to the perfect conditions for an explosion in mountain pine beetle populations. The insect now threatens the jack pines of the boreal (northern) forest, the largest biome on Earth, though the beetles' viability in those systems, and at those densities, is still unknown. As the climate warms, conditions for the beetles will only improve—an example of how climate change can destabilize formerly stable relationships within pristine ecosystems.

 PINE BEETLE, ACTUAL SIZE

Death by a Thousand Bites

For centuries the relationship was mutually beneficial: Pine beetles culled older, weaker trees, producing new beetles but also a healthier forest. Climate change, with its warmer, drier conditions, has upset that balance, leaving even healthy trees vulnerable to attack.

FIRST WEEK
Selection and Invasion
The cycle begins in summer, when a lone female beetle bores into a tree's bark and releases a pheromone that attracts hundreds of other beetles.

The tree tries to suffocate the insects by secreting resin into the beetles' boreholes.

SECOND WEEK
Burrowing and Egg Laying
Beetles dig galleries under the bark, depositing eggs and blue fungi to feed the next generation. The galleries block nutrient flow in the tree's phloem layer.

Sixty to eighty eggs are laid in each gallery.

Phloem layer

THIRD WEEK TO 4 MONTHS
Hatching and Feeding
Larvae hatch and chew side galleries, feeding on the phloem and the fungi.

◀ The tree remains green for months after beetles have fatally mauled it.

The larvae develop cold resistance in time for winter.

5 TO 12 MONTHS
Overwintering and Dispersal
The beetle larvae lie dormant until spring, when they'll turn into pupae, then adults. The new brood feeds on fungal spores before dispersing to another tree.

Pupal stage

◀ Needles turn yellow in the dry heat of summer.

Fungi-carrying new adult

13 TO 24 MONTHS
Red Means Dead
The beetles are long gone, and the drying tree turns red. Finally it loses most of its needles and becomes gray.

JOHN TOMANIO, NGM STAFF;
SHELLEY SPERRY
ART: SAMANTHA WELKER
SOURCE: DIANA SIX,
UNIVERSITY OF MONTANA

EXPLORE

IN THIS SECTION

A Beekeeper's Tools

The Inflated Charms of
Magnificent Frigatebirds

Polar Bear Selfies—
Lost and Found

ILLUMINATING THE MYSTERIES—AND WONDERS—ALL AROUND US EVERY DAY

TRIPLE THREATS

In Rocky Mountain National Park, 415 square miles of mountain terrain are protected—but not from effects of climate change. The average annual temperature in the high-elevation park increased 3.4°F in the 20th century. That has worsened a trifecta of troubles—bark beetles, wildfires, and invasive plants such as cheatgrass—doing visible harm to the plant life covering three-fourths of the park.

Colorado River

KAWUNEECHE VALLEY

Big Meadows
fire (2013)

34

Shadow Mountain
Lake

Grand
Lake
8,369 ft
2,551 m

Lake
Granby

Bighorn
Flats

ROCKY

MOUNTAIN

Chiefs Head Peak
13,579 ft
4,139 m

Longs Peak
14,259 ft
4,346 m

Bear
Lake
Road

Copeland
Mountain
13,176 ft
4,016 m

Ouzel
fire
(1978)

N. St. Vrain Cr.

WILD BASIN

7

Climate change leads to...

More cheatgrass

Increased wildfires

Bark beetle outbreaks

DOMINO EFFECT

Bark beetles, fires, and cheatgrass can play important ecological roles, but climate change exacerbates their effects on one another. For example, cheatgrass thrives when temperatures rise, adding kindling to wildfires that are already more intense due to drier conditions, and fires can spread faster where bark beetles have killed trees.

ATLAS BY **CLARE TRAINOR**

Mount Richthofen
12,940 ft
3,944 m

Source of the Colorado River

Long Draw Reservoir

Alpine Visitor Center

Trail Ridge

Ypsilon Mountain
13,514 ft
4,119 m

MUMMY RANGE

Hagues Peak
13,560 ft
4,133 m

N

SCALE VARIES IN THIS PERSPECTIVE. DISTANCE FROM ESTES PARK TO GRAND LAKE IS 18 MILES (29 KM).

Rocky Mountains
CO ROCKY MOUNTAIN N.P.
UNITED STATES

NATIONAL PARK

Trail Ridge Rd.

Old Fall River Rd.

Black Canyon Ck.

Fern Lake fire (2012)

34

36

Cow Creek fire (2010)

Rocky Mountain National Park Headquarters

Estes Park
7,522 ft
2,293 m

34

Projected suitable cheatgrass habitat in 2050

Current suitable cheatgrass (*Bromus tectorum*) habitat

Wildfire over 50 acres since 1970

Severe bark beetle damage, 2012–18

INVADING GRASSES

Non-native cheatgrass, once limited to the park's lowest elevations, is now spreading above 9,500 feet, moving more than 2,000 feet in elevation in just 10 years.

WILDER WILDFIRES

More acreage has burned here in the past eight years than in the previous century. The 2012 Fern Lake fire, caused by humans and fed by dry conditions, burned for months and over snow.

DESTRUCTIVE BEETLES

Dense stands of conifers are like a banquet for bark beetles. Mountain pine beetles infest 90 percent of the park's pine forests, and spruce beetle populations are rising fast.

MOUNTAINS: ISLANDS ON LAND

Mountains are complicated beasts. Not simply a mound of earth and rock, they are Earths in miniature. Mountains are warm at the bottom, but a short horizontal difference brings a big climatic change; moving uphill only 650 ft. (200 m) is equivalent to moving one degree toward the pole in latitude (temperature-wise, that's -2.3°F per 1000 feet (-4.2°C per 1000 m). This means that a hike of 25 mi. (40 km) from, say, Denver, Colorado (5250 ft. or 1600 m), to the peaks above town at 14,000 ft. (4260 m) is the equivalent of 13.5 degrees of latitude, the difference between central Texas and the US-Canadian border, or Rome and Sweden. Moving up in mountains means rapidly crossing climatic zones. As a result, mountains make for unique environments, popping up from the typical biogeography of a region to make islands in the sky.

Many mountaintops host endemic species found nowhere else, a direct result of their climatic isolation. Like island hopping in the Pacific, movement between mountaintops is fraught with challenges at best, downright impossible at worst. This is true regardless of the regional climate; the iconic snow-covered mountains of high latitudes are matched by tropical mountains that rarely or never see freezing temperatures and desert mountains that jut above dry and desolate plains. All are climatic islands. Perhaps the most biodiverse section of the inland North American continent, the Madrean Archipelago, is comprised of a series of sky islands jutting out of the southwestern US and northwestern Mexico, mixing multiple unique deserts and the subtropical Sierra Madre mountains with the cooler, more temperate Rocky Mountains of the north. Jaguars roam into Arizona, Chiricahua leopard frogs croak in the stream bottoms. Half of the world's biodiversity hotspots are in mountainous regions. This can be a blessing when thinking about the challenges of climate change; species can move up into cooler locations as the climate warms. But what happens when you reach the top? There is major concern about the survival of those species on mountaintops. As mountaintop environments disappear, so do the options they provide.

THE WORLD OF THE ROCKIES

Mountains have long been associated with myth, the abode of the gods: Olympus in Greece, Kailash (or Gang Rinpoche) in Tibet, Etna in Italy. The experience of being in mountains is one of overwhelming insignificance (not necessarily in a bad way!). They are a larger scale of perception, embodied in rock.

The story of mountains is written in the rocks, and best appreciated when the topography is united with the geology, as this piece illustrates, using modern elevational data and a 1916 geology map.

GEOLOGICAL MAP
OF THE
STATE OF CALIFORNIA
ISSUED BY
State Mining Bureau
COMPILED UNDER THE DIRECTION OF
FLETCHER HAMILTON, STATE MINERALOGIST.
GEOLOGY BY JAMES PERRIN SMITH.

1916.
REPRINTED 1929

> … mountains, like all wildernesses, challenge our complacent conviction—so easy to lapse into—that the world has been made for humans by humans. Most of us exist for most of the time in worlds which are humanly arranged, themed and controlled. One forgets that there are environments which do not respond to the flick of a switch or the twist of a dial, and which have their own rhythms and orders of existence. Mountains correct this amnesia.
>
> —Robert MacFarlane, *Mountains of the Mind*

Perhaps the oldest map ever discovered is engraved on the face of a mammoth tusk, found in the Czech Republic (circa 25,000 years BCE). It appears to depict the landscape around the Dyje River, along with routes around the hills and mountains edging the river valley. Calling early etchings maps, or representations of geographic space, is fraught with difficulty. But it is unsurprising that these massive features of the landscape would be the subject of early charting. Mountains dominate the mind and the landscape alike. Charting your way around them was, and is, an important task for scientists, cartographers, and anybody trekking the landscape. But let's consider a more recent human experience with mountains.

As you approach the US Rocky Mountains from the east, you first must traverse the Great Plains, the prairies and steppes of North America. This is flat land, imperceptibly rising from the lowlands of the Mississippi River over thousands of kilometers. Often dry, certainly monotonous—but doable. Thousands did exactly that in the famous covered wagons of the 1800s. But slowly, as the land dried further and the grasses shrank to stubble, a long shadow appeared in the west. These were the Rockies, rising like a wall running shockingly straight north and south, in places shooting up 1000 ft. in just half a mile (300 m in 0.80 km). It is unfair to call them foothills; that implies a sort of gradual ramp up to the "real" mountains. No, to the settlers, the Rockies must have seemed to jump out of the ground suddenly and unforgivingly, like castle ramparts erected against their puny wagon trains. The Rockies are a high (14,000 ft. or 4300 m), long (2000 mi. or 3100 km), and wide range, spread out over 600 mi. (965 km) east to west. Their formation around 75 million years ago was the result of several shallow plates sliding under the North American continent. But as they went under, they were not steeply pitched down into the planet, as many other ranges. The angle was shallow, meaning the uplift extended farther inland. The mountains arose in a series of blocks and basins, with fractures driving mountains up and basins down along fault lines—hence, the steep topography. Further glaciations and erosion created the incredibly precipitous, blocky,

In 1837, Benjamin Bonneville (who named Lake Bonneville, the remnants of which are now the Great Salt Lake) mapped the mountain ranges and tribal cultures of the western United States. Several archaic spellings and place names are visible here, from the eponymous Lake Bonneville to Clark's River (now the upper Columbia River, visible in the upper left of the map).

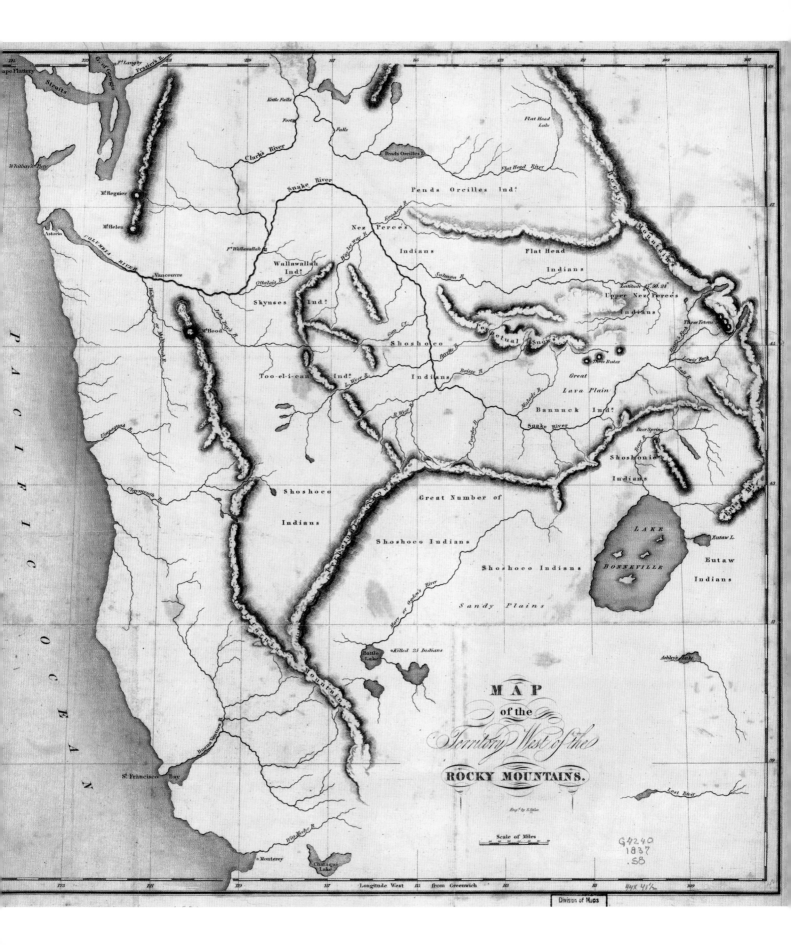

MAP
of the
Territory West of the
ROCKY MOUNTAINS.

Engd by S. Stiles

Scale of Miles

VANCOUVER'S ISLAND

STRAIT OF JUAN DE FUCA

Cape Flattery

P A C I F I C

O C E A N

Mt Olympus

Cape Disappointment

Point Adams

Fort Langley

Mt Baker

Whatcom

COLUMBIA RIVER

GREAT PLAIN

Mt St Helens

Mt Adams

Ft Wallah-Wallah

C O L U M B I A   R I V E R

(Supplementary Sketch)

Reconnaissance of the Railroad Route

From WALLAWALLA to SEATTLE via

YAK-E-MAH RIVER & SNOQUALMIE PASS.

By A.W. Tinkham.

Scale 10 Miles

Fort Wallawalla

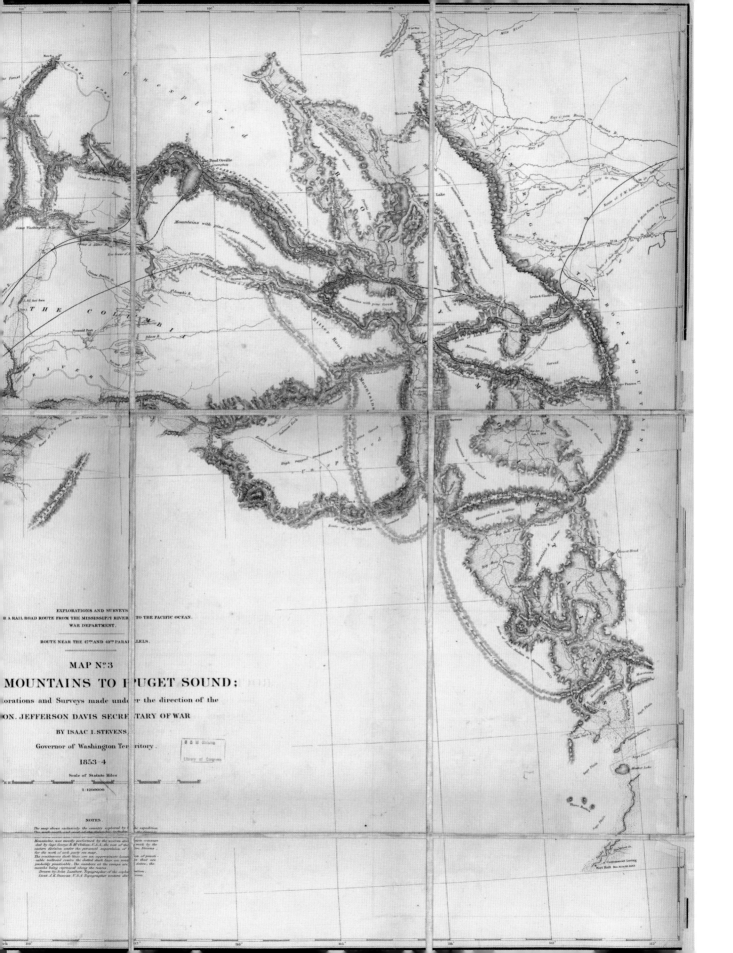

EXPLORATIONS AND SURVEYS
B A RAIL ROAD ROUTE FROM THE MISSISSIPPI RIVER TO THE PACIFIC OCEAN.
WAR DEPARTMENT.

ROUTE NEAR THE 47TH AND 49TH PARALLELS.

MAP Nº 3

MOUNTAINS TO PUGET SOUND;

orations and Surveys made under the direction of the

ON. JEFFERSON DAVIS SECRETARY OF WAR

BY ISAAC I. STEVENS,

Governor of Washington Territory.

1853-4

Scale of Statute Miles

1:1200000

NOTES.

The map shows exclusively the country explored by the expedition.

Mountains, was mostly performed by the western division...
ded by Capt George B. Mc Clellan U.S.A., the rest of the
eastern division, under the personal superintendence of Gov. Stevens
for the work of each party see map.

The continuous dash lines are an approximate location of pract-
cable railroad routes the dotted dark line are roads so that are
probably practicable. The numbers at the camps are dates, the
months being expressed along the routes.

Drawn by John Lambert, Topographer of the exploration.
Lieut. J.K. Duncan. U.S.A. Topographer western division.

and cracked landscape, full of streams and boulders which harbored people and cultures for thousands of years and deeply impressed early European explorers with their beauty and ruggedness. The explorers' maps show the incredible impact of mountains on life, whether that life establishes in the mountains or seeks to pass through them.

> This southern escarpment [of the Tavaputs Plateau, a ridge in the southern Rocky Mountains] presents one of the most wonderful facades of the world. It is from 2,000 to 4,000 feet high. The descent is not made by one bold step, for it is cut by canyons and cliffs. It is a zone several miles in width which is a vast labyrinth of canyons, cliffs, buttes, pinnacles, minarets, and attached rocks of Cyclopean magnitude, the whole destitute of soil and vegetation, colored in many brilliant tones and tints, and carved in many weird forms—a land of desolation, dedicated forever to the geologist in the artist, where civilization can find no resting-place.
>
> —J. W. Powell, geologist and explorer, *The Exploration of the Colorado River and Its Canyons*

EXPANDING LIFE AND RETREATING GLACIERS

Just as mountains are miniature Earths, their climates are changing, too. As temperatures rise, the climatic zones defined by elevation creep uphill, pushing colder zones off the top and allowing new species to invade at the bottom.

Some life is expanding. As snow disappears, the growing season at high elevations gets longer, allowing trees to invade former subalpine grasslands. High-elevation tree line is a somewhat difficult concept to define scientifically, but it is easily recognizable to anybody venturing to high-altitude summits. It is worth a brief note to say that tree line at any given location is determined largely by local factors—things like soil, solar exposure, and local topography. At a fine scale, things get complicated. But globally, there is a strong link between tree line and climate, as mentioned. And fully half of mountain tree lines around the world are moving uphill, evidence of the warmer atmosphere (where tree line fails to move uphill, it is often a result of shrubs invading first and keeping the trees out). As we've seen throughout this book, life expands into opening landscapes—from Humboldt's Chimborazo research to Glacier Bay, Alaska. As land frees up, either climatically or literally (as when a glacier melts), life will come.

PREVIOUS
John Lambert, topographer, created this 1859 map of potential railroad routes (dark lines) through the Rocky Mountains and the narrow but steep Cascade Range of mountains in Washington and the incised canyons of the Columbia River, both shown in fuzzy, shaded relief. A supplementary map of the key pass from "Wallawalla [now Walla Walla] to Seattle via Yak-e-ma River [now Yakima] and Snoqualmie Pass" is from A. W. Tinkham (1854).

RIGHT
The invention of radar imagery enabled the mapping of rocky features below the vegetated surface, revealing faults, cracks, lava flows, and other geologic features in stark relief. This map, from 1968, was one of the first. It highlights the complex geological beauty of Yellowstone National Park; the circles show joints, faults, and lava flows associated with complex ground forces that shape the hot springs, mountains, and geysers of the geologically active region.

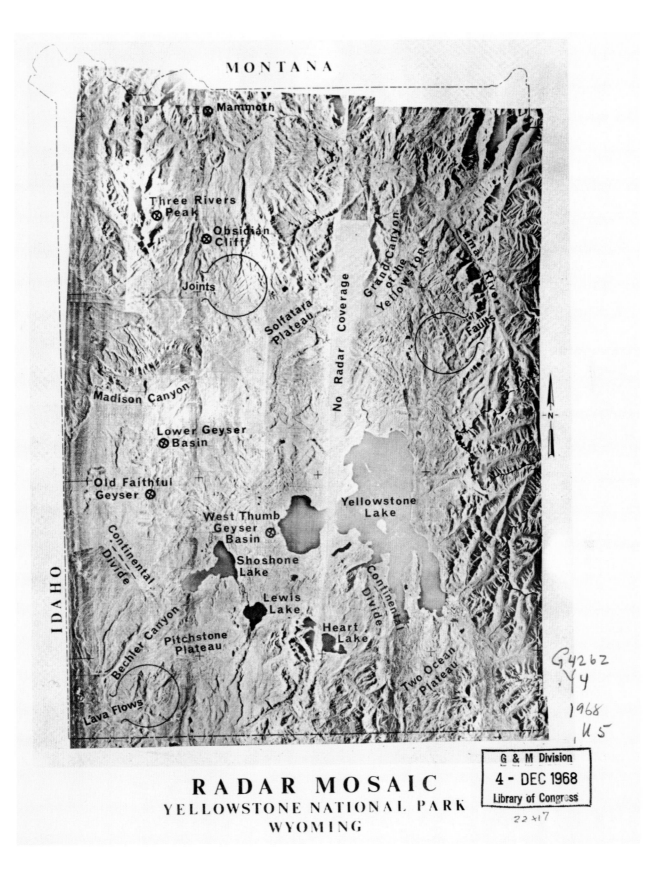

MONTANA

Mammoth

Three Rivers Peak

Obsidian Cliff

Joints

Solfatara Plateau

No Radar Coverage

Grand Canyon of the Yellowstone

Lamar River

Faults

Madison Canyon

Lower Geyser Basin

Old Faithful Geyser

West Thumb Geyser Basin

Yellowstone Lake

Shoshone Lake

Continental Divide

Lewis Lake

Continental Divide

Heart Lake

Bechler Canyon

Pitchstone Plateau

Two Ocean Plateau

Lava Flows

IDAHO

N

RADAR MOSAIC
YELLOWSTONE NATIONAL PARK
WYOMING

But when there is limited space, as one thing expands another generally retreats. The big losers in mountain systems are glaciers and alpine environments. The climatic zones that foster those systems simply move off the top of the mountain. Glaciers start to disappear when they run what is termed a "negative mass balance," which is literally the balance between incoming mass (snow) and outgoing mass (melt). A negative balance means more is leaving then coming in, averaged over the year. Glaciers are constantly melting and flowing; after all, they are rivers of ice. Typically all that outflow is matched by precipitation coming in at the top, snow which is compressed into ice. A negative mass balance results in a shrinking glacier—either retreating at the bottom (with the front starting to retreat as it melts faster than it flows), shrinking on the sides and top as it thins, or both. Since 1961, mountain glaciers (excluding Greenland and Antarctica) have shed more than 9000 billion tons of ice: an ice cube the size of Montana, nearly 100 ft. (30 m) tall. (To connect to the previous conversations, that's enough to raise sea level about 1 in. or 27 mm.) This negative mass balance is a result of mountain areas warming about 0.5°F (0.3°C) per decade. In some places, most of this warming is in the summer (the Alps), in others mostly the winter (Tibet). The highest mountains may see increased snowfall, as extreme elevations will likely remain below freezing despite warming, but it seems unlikely that such snowfall will offset ice losses at lower elevations. (Some regional areas, such as the Karakoram mountains on the Asian continent, currently appear to be stable. This seems likely a result of local factors such as increased cloud cover slowing melt and more snow in the past decades.) But regardless of our fine-scaled human experiences on this or that mountain, with this or that glacial system, and in this or that year, the global glacial system is slowly trickling away into the seas.

Historically, glaciers and high mountains have been the water towers of their landscapes, poking their peaks up into the river of circulating atmospheric moisture and collecting snow and rain to nurture the landscapes below them. The snow is stored and then metered out over the summer, providing a base flow when the land needs it most. But as more snow falls as rain, and the snow that

Like its application in rivers, LiDAR can be utilized in geological settings as well, such as the glaciers on Washington State's Mount Rainier shown here. Historical landslides, scarps, and fault lines are revealed via subtle variations in elevation; color can then be used to highlight the patterns.

does fall melts 5 days sooner per decade on average, that water capture, storage, and distribution system is breaking down. Even if total moisture doesn't change—and in some places it is going up as glaciers melt faster—late summer can run dry as the melt pushes out sooner and in a more concentrated pulse. In developing countries like Nepal, this has led to a direct loss of high production crops such as rice, which takes considerable amounts of water, meaning calorie shortages, hunger, and a loss of income. Sometimes, farms are simply abandoned. In the American West, a dry land nurtured almost entirely by mountain snow, spring snow cover has declined about 20 percent since 1980 (spring, specifically April 1, is typically the high point for snow cover and is used for comparing between years). Mountain snow there provides 75 percent of the water for upward of 70 million people, and that snow is disappearing rapidly—since 1915, the total average mountain snowpack there has declined 15–30 percent. The future functioning of the water towers is uncertain, with thirsty landscapes waiting below.

GRASSLANDS: SAVANNAS AND STEPPES

Take the precipitation in a forested landscape a notch down and you get a grasslands region, a landscape too dry for trees (annually or seasonally, and fire can play a role) but too wet to be considered a desert. Steppes, from the Russian term for "flat, grassy plain," are dry, cold, seasonal grasslands, oceans of perennial and annual grasses and wildflowers far from the ocean. In the tropics, you have savannas. These two landscapes are distinct in some ways but merge together in others. Steppes are colder, with fewer trees, higher variations in temperature, and typically lower rainfall; some go further and divide steppes into true steppes (shortgrasses) and prairies (tallgrasses). Savannas are wetter and warmer, typically with trees at relatively low densities. However, some tropical savannas have tree densities approaching forest levels, though they are smaller and more evenly spaced. Both are grasslands, and depending on how you define grasslands, they comprise as much as 20–40 percent of the world's total land. (Clearly, partitioning the world into tidy categories for a map is a messy exercise at best, and every little local region has its own flavors.)

THE DOMINANT FEATURES OF THESE STEPPES AND SAVANNAS, THE ICONIC OPEN SPACES OF OUR PLANET, ARE GRASS AND OPEN VISTAS.

The dominant features of these steppes and savannas, the iconic open spaces of our planet, are grass and open vistas. The largest examples are the transcontinental steppes of Ukraine, Russia, Kazakhstan, and the other central Asian countries—true wonders of the world and a highway for horsemen in the distant past.

The tallgrass prairie flowers of North America, and the birds and insects that call this land home.

The Great Plains of the United States, Canada, and Mexico are another example, as is the windswept desolation of Patagonia in Argentina and Chile. Their topographic simplicity belies their belowground bounty. Thick, rich soil develops under the grasses as they die and slowly decompose, building the agricultural breadbaskets of continents. Patches of similar vegetation exist elsewhere as well, in the Mediterranean basin and Australia. In the tropics, the well-known savannas of Africa (covering about a third of the continent, depending on where you draw the lines) beckon photographers and tourists alike, while locals vie with the local wildlife for crops and water.

The open vistas and big skies are clear and blue, and the wildflower diversity is stunning. Even in temperate latitudes it can get quite dry of course, often in winter; summer provides the occasional wet thunderstorm that replenishes the land.

Nevertheless, the precipitation on average is too low or too seasonal for most trees, and the occasional extreme drought plus fire and ample grazing animals generally keep the trees limited to riverside galleries.

Tropical grasslands such as the savannas typically get considerably more water on an annual basis than their cousins at higher latitude. Here precipitation seasonality is more important than temperature seasonality; grasslands are maintained by a long and intense dry season, in which fires are common. These fires eliminate any poor tree unlucky enough to get started in the grass, which is unperturbed by a little burning. The same amount of rain distributed throughout the year would likely result in a forest; it is the seasonal dryness that enables those fires. (Sometimes, small patches of grassy savannas result from soil conditions that get waterlogged in the rain, meaning trees effectively drown. They are not as common but certainly a feature of the landscape, especially in Africa). Drought is the key feature, whether annual or seasonal. Grasses like it dry and can take the heat.

THE STORY OF THE GREAT PLAINS

The Great Plains of North America offer prime examples of both prairies in the east and cold, shortgrass steppes in the west. (Tropical Mexico also has some proper savannas—known as *sabanas* or *pastizal de clima caliente*—grasslands with rainfall in excess of 40 in. [1000 mm] a year, but a well-defined dry season and fires. These are quite limited in extent, less than one percent of the country.) Today, many might visualize the Great Plains as flyover country or the long, boring stretch of highway on a road trip, a featureless place devoid of character. It wasn't always so. Early records note the enormous sky, the gentle, varied topography, the ample wildlife, the luxuriant grasses, and flashy wildflowers. Any dearth of beauty today is our responsibility:

> ... we stood by and allowed what happened to the Great Plains a century ago, the destruction of one of the ecological wonders of the world.... we need to see this with clear eyes, and soberly, so that we understand well that the flyover country of our own time derives much of its forgettability from being a slate wiped almost clean of its original figures.
>
> —Dan Flores, *American Serengeti*

It still is a great ecological wonder in places, if you're lucky enough to find some intact tallgrass or experience the shortgrass steppe in the early spring. The Great Plains offered tremendous bounty for thousands of years to the animals and

The Missouri Plateau, the western edge of the Great Plains, as seen from the International Space Station by the Expedition 57 crew, looking from above the Rocky Mountains toward the east. Snowy mountaintops supply water to a band of dark forests at middle elevations, but the lower expanse—much drier—is a sea of grasses.

MISSOURI PLATEAU

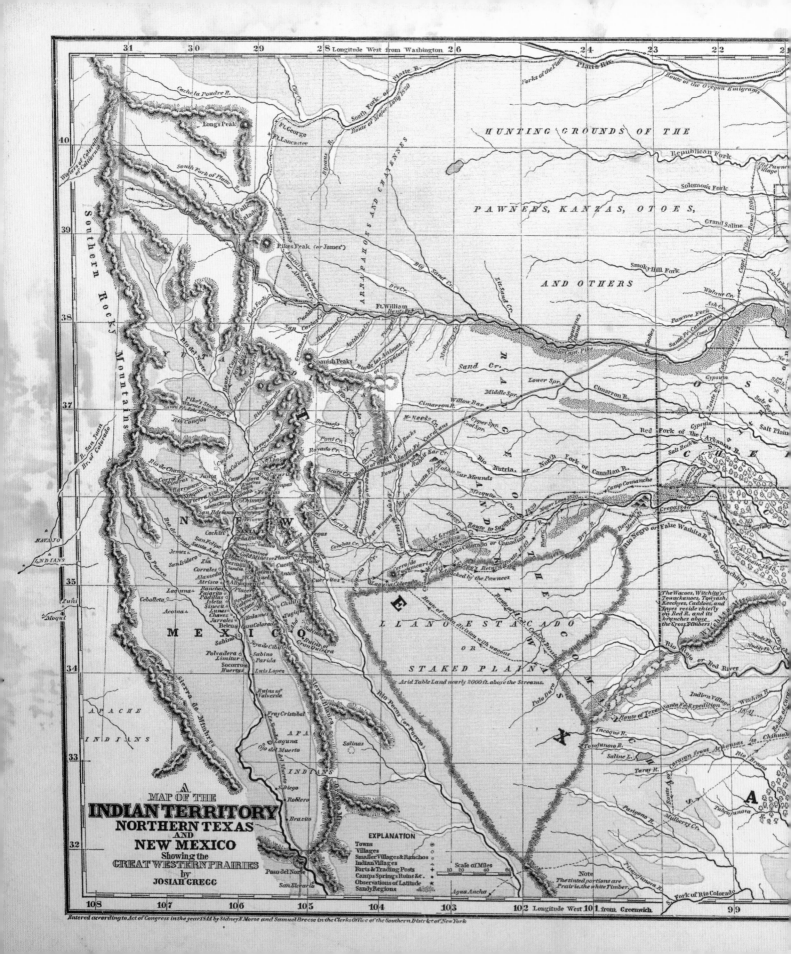

A
MAP OF THE
INDIAN TERRITORY
NORTHERN TEXAS
AND
NEW MEXICO
Showing the
GREAT WESTERN PRAIRIES
by
JOSIAH GREGG

EXPLANATION
Towns
Villages
Smaller Villages & Ranchos
Indian Villages
Forts & Trading Posts
Camps Springs Ruins &c.
Observations of Latitude
Sandy Regions

Scale of Miles
10   20   40   60

Note
The tinted portions are
Prairie, the white Timber.

Native Land Use

Far from a desolate, grassy sea, the Great Plains of the United States were populated by numerous Indigenous peoples prior to European expansion, as this 1844 map by Santa Fe trader Josiah Gregg shows. The extensive notes show the reliance of people on waterways and the rare timberlands, as well as the geometrically precise but ecologically foolish resettlement areas for displaced eastern tribes (rectangular sections seen on the right), which display a European habit of breaking the land into square plots and edges without regard for ecosystems or climatic constraints on life.

people that frequented the region's grassy shores. Immense herds of bison migrated throughout the land, and the Indigenous communities followed, tethered to this magical source of food and material. Around 1100 CE, in the wetter parts of the Plains, a thriving agricultural practice developed, a fundamental achievement and massive change that supported vast Indigenous civilizations. The reintroduction of the horse in the mid 1700s was a second fundamental change. Horses, gone for about 10,000 years in North America, arrived via European colonizers and quickly revolutionized life on the Plains. The grassy sea could be utilized more directly. Buffalo herds could be followed in real time. Humans could fully exploit the sporadic nature of the grassland ecosystem, a land ruled by variability in rainfall and by movement.

HORSES ARRIVED VIA EUROPEAN COLONIZERS AND QUICKLY REVOLUTIONIZED LIFE ON THE PLAINS.

European settlers encountered the same ecosystem but took a decidedly different approach to the natural world, one that attempted to erase the variability rather than move with it.

A bit more geography. The North American Great Plains run from about 95 degrees to 105 degrees or 110 degrees west longitude, depending on how far north or south you are. You can roughly divide the region by the 100th meridian, the line that runs north and south through Kansas and into Manitoba (immortalized as the place "where the Great Plains begin" by the Tragically Hip, a Canadian rock band). East of the 100th meridian, the climate is wetter, generally controlled by warm waters from the Gulf of Mexico. But to the west, the drier steppe landscape was "almost wholly unfit for cultivation and, of course, uninhabitable by people depending upon agriculture for their subsistence," noted Edwin James in 1823. In the east, agriculture thrived—this was the same landscape that Indigenous communities developed into rich agricultural civilizations before European contact. Growing maize and local species of barley, Indigenous agriculture supported immense and complex societies. As the weather varied, agriculture would advance westward (in wet years) or contract eastward (in dry years) as the people flexibly adapted their food sources to the climate.

Europeans approached this same variability in the opposite way. Although a defining characteristic of grasslands is precipitation variance, and water is the primary constraint on agriculture, they sought to wrest water from the skies and ground when the weather did not provide. The vast plains were too great and the intoxicating power of wet years too strong for a young and ambitious country. The years when the Plains were green from end to end seduced people into thinking agriculture was possible anywhere with enough willpower, natural variability be damned. The land was duly divvied up into rectangular blocks for permanent homesteads and farms—nice, straight-edged geometric patterns inscribed on a highly variable ecosystems. Some concessions were made; west of the 100th meridian, wheat replaced corn as it required less water, farms were bigger to maintain profitability. This is apparent even in the basic infrastructure and populations that live in the region today—a sudden drop in population as you move west, due to farms getting larger to accommodate the lower productivity per unit area. But while there are fewer total people, the regularity of the farm economy was still imposed on the near-entirety of the prairie. After all, the notion of adapting to dry years via movement was a non-starter, as private property meant you owned the land under your feet, for better or worse. So rather than work with the variability, they forced regularity onto the landscape itself. Nature follows geography, but the European settlers followed geometry—nice, square, evenly spaced farms—much to their detriment.

MINING FOR WATER

When the southern Rocky Mountains were being built, the forces of erosion were just as active as they are today. Rising mountains met ice and rain, cracking the ancient sandstone and sending an immense amount of coarse sand and gravel down creeks and gullies, out into the nascent Great Plains. For several million years, the rough sediment sloughed across the western, arid portion of the region, downslope from the mountains (the shortgrass steppes, the region least blessed with rain), in some places nearing 1000 ft. (300 m) in depth. In paleo-valleys where the runoff collected it was even deeper. That coarse material came in handy later, specifically the gaps in between those coarse grains. It was (and is) able to hold a considerable amount of water in those porous spaces. As the last Ice Age ended, water stored over thousands of years in glacial ice made its way down the High Plains and into the thick layers of sand and gravel, now deeply covered by the steppes and grasslands we know today—enough water to cover the *entire* United States about 1.5 ft. (0.5 m) deep. Called the Ogallala Aquifer after a nearby town (also referred to as the High Plains aquifer system, as there are several aquifers in the region), this immense layer of soaked sand was just out of reach below the grasses and shrubs that make up the steppe. It moves an estimated 1 ft. (0.3 m) per day to the east, sloping down from the mountains, and in this arid landscape takes about 6000 years to replenish. Using it is akin to mining a nonrenewable resource; the waters of the Ogallala and the other aquifers are fossil waters, old and slow.

FOLLOWING
An 1843 map from the US Congress shows the international border with Mexico that was soon to be moved; land south would belong to the United States within the decade. (Horizontal border runs east–west on lower half of map.)

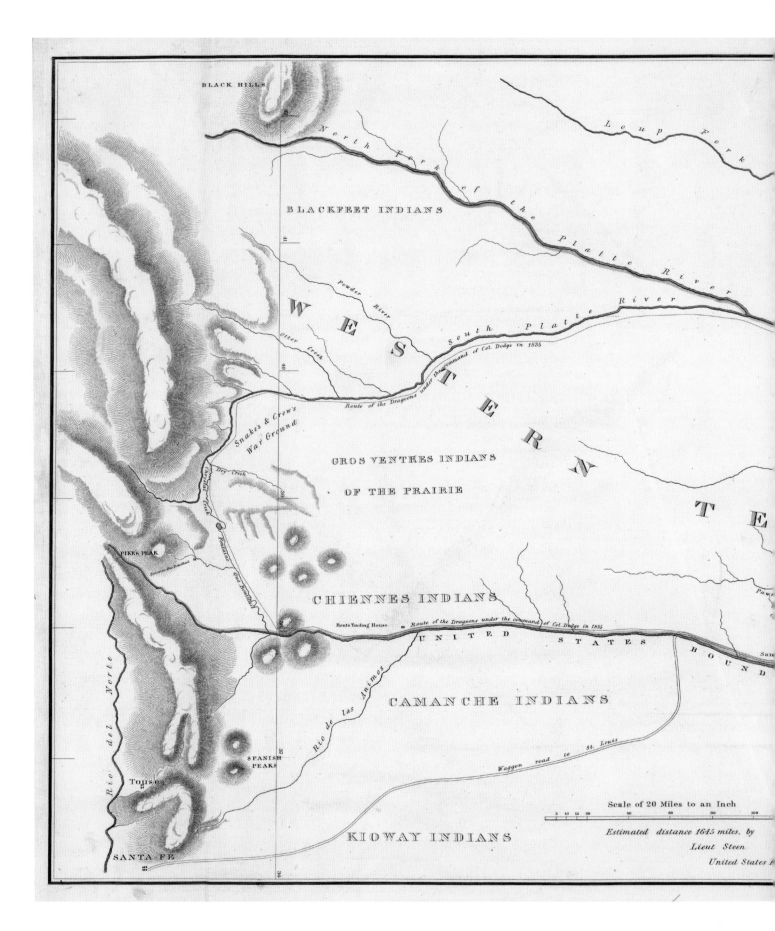

BLACK HILLS

BLACKFEET INDIANS

North Fork of the Platte River

Loup Fork

W E S T E R N   T E

Powder River

Otter Creek

South Platte River

Route of the Dragoons under the command of Col. Dodge in 1835

Snakes & Crow's War Ground

GROS VENTRES INDIANS

OF THE PRAIRIE

Dry Creek

Chalate Creek

Fontaine Qui Bouillit

PIKES PEAK

CHIENNES INDIANS

Bents Trading House   Route of the Dragoons under the command of Col. Dodge in 1835

U N I T E D   S T A T E S   B O U N D

Rio del Norte

CAMANCHE INDIANS

Rio de las Animas

SPANISH PEAKS

Tousse

Waggon road to St. Louis

SANTA FE

KIOWAY INDIANS

Scale of 20 Miles to an Inch

5 10 15 20   40   60   80   100

Estimated distance 1645 miles, by

Lieut Steen

United States

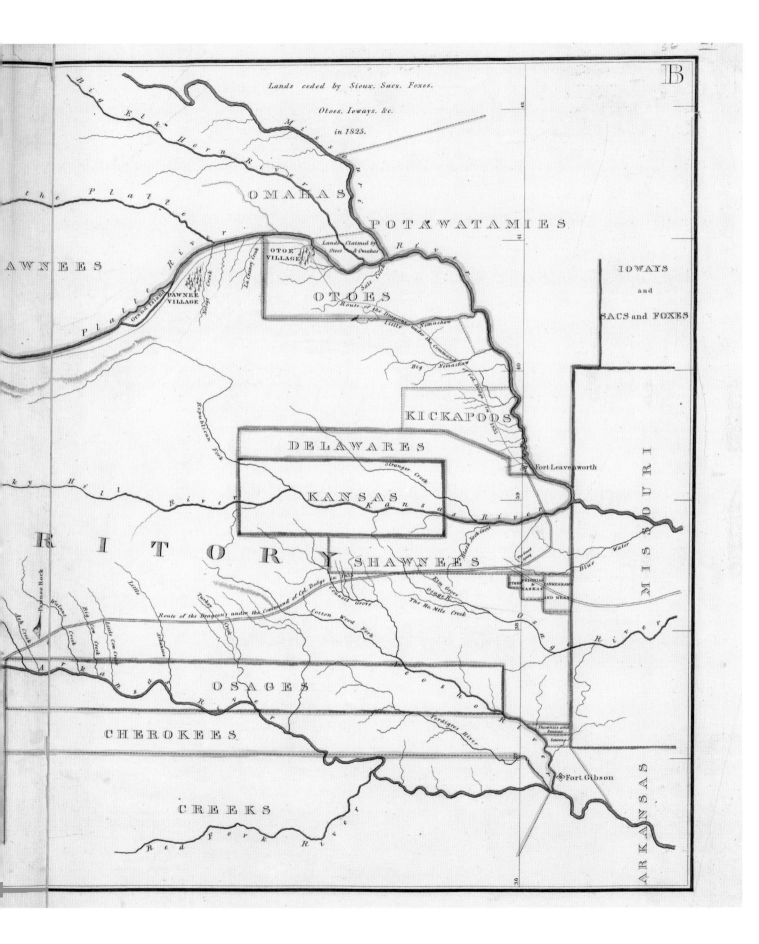

That ability came in handy on the 19th-century Plains. Remember, the farther you move west of the 100th meridian, the less annual rainfall. But there were good years and bad. Farmers intent on scratching out an existence on their own land were entranced by the good years and determined to outwit the bad, forcing regularity on an irregular system.

They finally did so successfully by tapping that immense natural water source just underneath their plows. The discovery did not come out of nowhere. Windmills had been installed for livestock throughout the Great Plains for decades, pulling up water from the mysterious source below. Water was so close to the surface in parts of Texas that seemingly unlimited wells could be dug easily by hand. Finally, N. H. Darton, a geologist, determined in 1898 that all the wells in his portion of Nebraska (near the town of Ogallala), plus the wells in Colorado and Kansas, were drawing water from the same geological formation. The vast network of canyons, gullies, flats, and burrows—a network buried by Rocky Mountain sand and gravel and then flooded and buried again—was given its name.

Pumping began immediately but was quite constrained in scope by the limitations of windmill technology. Only a few years before the Dust Bowl began in 1930 (when the prairies began to blow away after years of farming abuse left them without a vegetation covering during a drought), the Nebraska Agricultural Extension (NAE) said that while "the underground water supply is abundant," technology limited the farmers' abilities in "lifting it to the surface and applying it to the land." During the Dust Bowl, farmers watched their fields literally fly away for lack of rain, despite the massive freshwater resource sitting under their feet. Kansas soil took to the air, the product of centuries often disappearing in an afternoon. Dust fell on Washington DC and into the Atlantic Ocean.

But the NAE soon got its wish for stabilizing water. The invention and perfection of the internal combustion engine and the mass industrialization of the United States during World War II brought fossil fuels to bear on the main problem with this fossil water: its inaccessibility despite its tantalizing proximity. And once the problem of extraction was solved, farmers wasted no time in taking advantage of the vast subterranean resource. Together with ample sunshine and the heat of the plains, abundant water finally created the perfect recipe for agricultural success. From the 1950s to the 1970s, water withdrawals increased 500 percent.

Sadly, the Ogallala Aquifer was not inexhaustible, and only replaceable on a timeline measured in centuries. By 1980, water levels had dropped 10 ft. (3 m) across the 600-mile region. Today, entire parts of the Ogallala reservoir are tapped

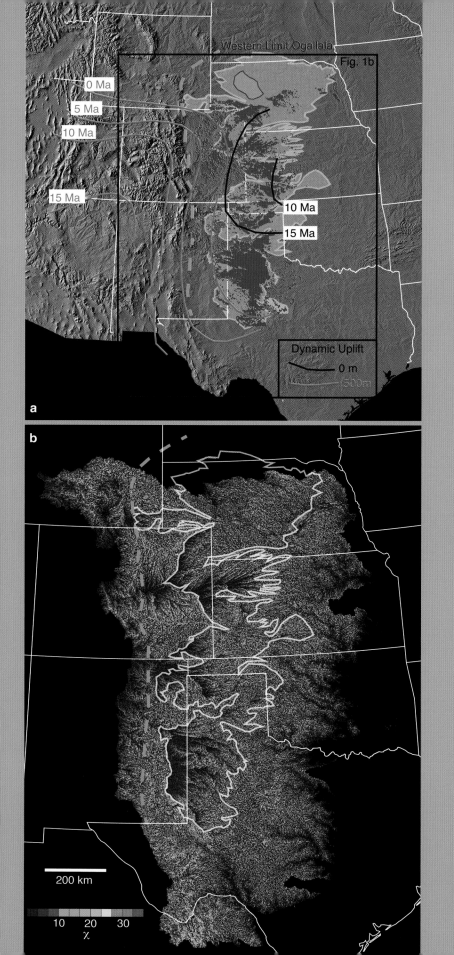

Ogallala Aquifer

The Ogallala Aquifer and the rocks underlying the surrounding plains are comprised of sediments eroded from the Rocky Mountains, which sloughed onto the plains and filled paleo-historical river channels; those sediments hold considerable water. This massive geological event is still unfolding as drainages continue to evolve and change. The top image shows the aquifer in its past context; red lines indicate the approximate topography at various points millions of years ago (Ma), where the land was moving upward; black lines are the end of that uplift, the "pivot point" for elevational change. The dark blue areas indicate likely historical lakes. In the bottom image, created from a topographical model, areas that are still forming mature drainage networks are highlighted in red, indicating instability (high values). Areas that are more stable are in blue (low values). The solid white line indicates the traditional Ogallala boundaries.

In all, these subsurface images and maps reveal how the vast water resource formed and how it continues to evolve under our feet, slowly, even as we deplete it.

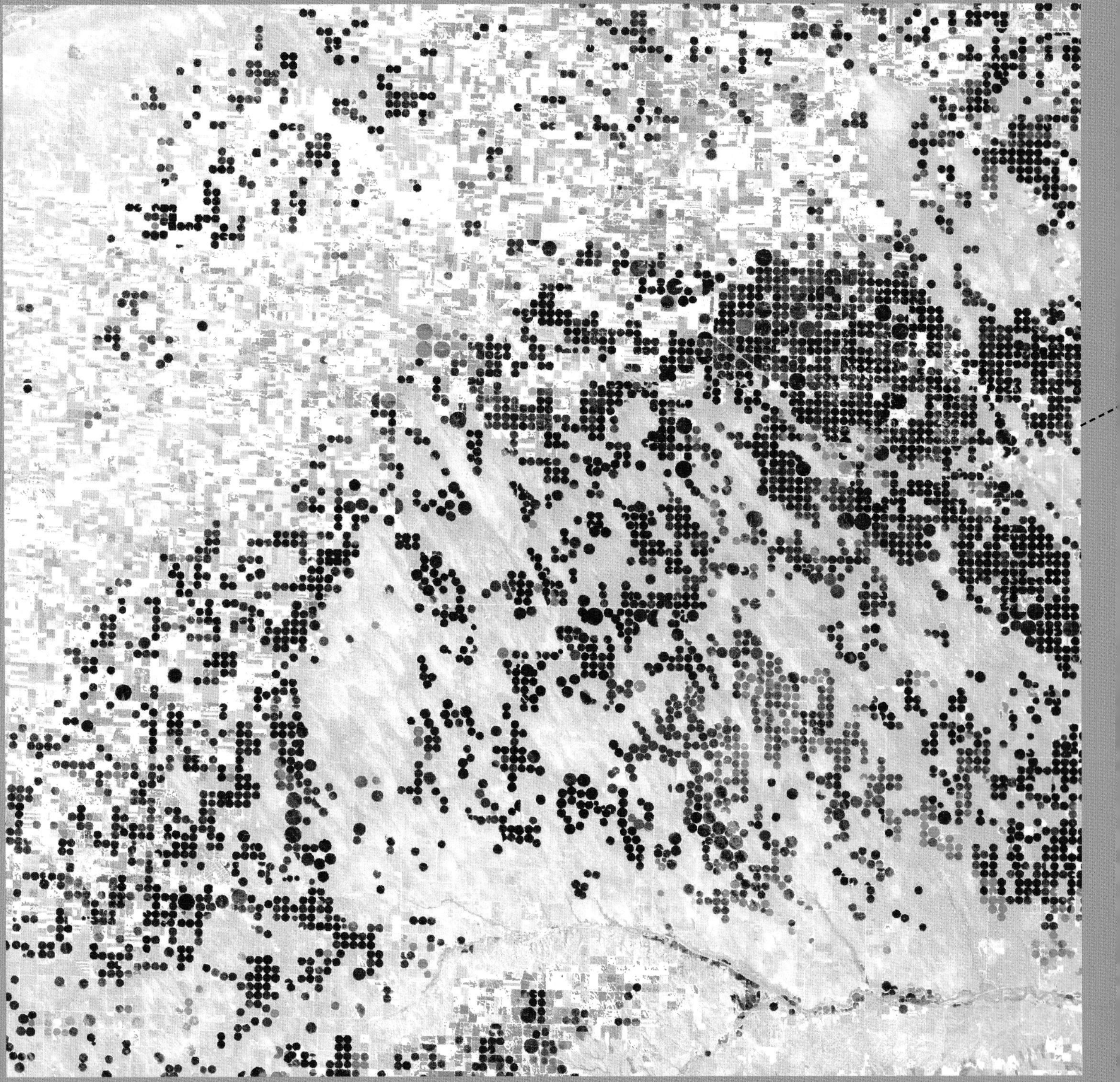

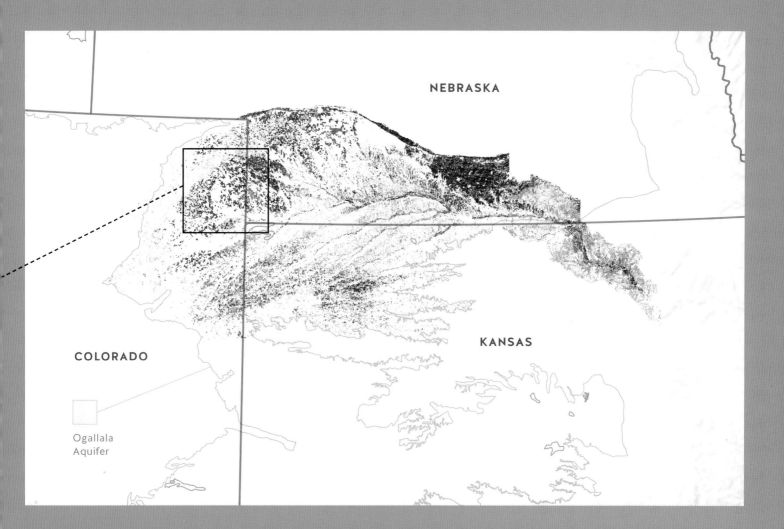

NEBRASKA

COLORADO

KANSAS

Ogallala
Aquifer

Irrigation Patterns

To stabilize water supplies between dry and wet years, farms pump water from
the ground. Some fields are watered almost every year, others are more sporadic as a result of differing crop types, weather, or farmland expansion and
contraction. The circular shapes of the fields result from central pivot irrigation systems, giant sprinklers that rotate around a central pipe. These form a
pattern of crops reminiscent of the pointillism painters of the late 1800s. The
images here, from the borders of Nebraska, Kansas, and Colorado and spanning from 1999 to 2016, show the number of years irrigated over that period
on a field-by-field basis. Colors indicate continuous irrigation, from 0 to 18
years. Dark blue shows fields irrigated every year. The lighter the colors, the
less frequently the fields were irrigated.

out, with no water remaining. Farmlands can draw down the aquifer's water at a rate of 4–6 ft. (1.5–1.8 m) per year, while natural replenishment adds only half an inch. Some areas—where the climate recharges the aquifer faster and the need for additional irrigation is less—are marginally sustainable. The Ogallala is a vast water battery that still powers the Plains during periods of drought. But it is a battery that is running low.

CLIMATE CHANGE AND GRASSLANDS

Earth's grasslands occur in delicate climate spaces: either that narrow dryness that supports grass but not trees, persistent cool weather, or the variable climate in which extreme droughts preclude trees and shrubs via water stress or fire (and in some cases, grazing). These lands also have the unlucky blessing of being incredibly fertile—soils that can feed the world—and so they are coveted by humanity. Their fate in a changing climate and an increasingly dense human world is very uncertain, perhaps more so than other systems in this book, because everything hinges on that variance.

Over a century ago, the 100th meridian was a signpost emphasizing the constraints that natural variability in climate places on human endeavors. The sharp shift in life was easily noticed. In 1890, John Wesley Powell wrote:

> On the east a luxuriant growth of grass is seen, and the gaudy flowers of the order Compositae [flowers such as sunflowers, daisies, and asters] make the prairie landscape beautiful. Passing westward, species after species of luxuriant grass and brilliant flowering plants disappear; the ground gradually becomes naked, with 'bunch' grasses here and there; now and then a thorny cactus is seen, and the yucca plant thrusts out its sharp bayonets. At the western margin of the zone the arid lands proper are reached.... In seasons of plenty, rich crops can be raised without irrigation. In seasons of drought, the fields are desert.

But that signpost is moving.

The conditions described by the 100th meridian, the sharp decline in rainfall that left its imprint on crop patterns, farm sizes, and human densities, is shifting east as the climate warms. In the United States, the shortgrass steppes are drying, as are the wetter prairies. The effective 100th meridian will approach the real 95th meridian in the near future. It is a symbol of how modern agriculture and humans interact with the environment—where variability in the grassland system can be tolerated (the wetter east) and where it must be forcefully limited (the drier east). In areas that are getting wetter, trees are encroaching and threatening the native biodiversity. Around 50 percent of temperate grasslands are permanently lost to agriculture, and around 15 percent of tropical savannas are similarly gone. Only about 10 percent of the world's grasslands are protected. The future of the wild grasslands of our planet is very much in doubt, as much due to human pressures for food as climate change.

DESERTS: LIFE WILL FIND A WAY

Stepping further up the aridity gradient and down in moisture, we arrive at deserts, the low point of the precipitation scale. Deserts are defined by two criteria: terrestrial status and low precipitation. The first is straightforward. There are certainly parts of the ocean which get extremely little rain, but nobody would call them deserts. The second is precipitation, or rather the lack thereof. A common cutoff is 10 in. (25 cm) of rain or less annually. At that threshold, about a third of Earth's land surface is desert.

BOTH HOT AND COLD DESERTS ARE QUITE COMMON.

Note that there is no temperature threshold in our conception of deserts; both hot and cold deserts are quite common. The hottest temperature recorded in a desert (and the highest ever recorded in the world) was in Death Valley, California, on July 10, 1913: 134.1°F or 56.7°C. The highest average annual temperature is also in a desert, the little-known Danakil Desert, located in northern Ethiopia. Over the period from 1960 to 1966, the verified annual average temperature there, including both day and night, was 94.0°F or 34.4°C! These two record-holding locations have one distinct advantage—they are well below sea level, which increases air temperatures as a result of higher atmospheric pressures. But Antarctica is also a desert, a high plateau of ice—yet with little precipitation.

Once past the precipitation threshold, deserts come in a variety of flavors. The blowing dunes and rocky outcrops of the Sahara (3.6 million sq. mi. or 9.4 million sq km) and the snowfields of Antarctica (5.5 million sq. mi. or 14.2 million sq km) are the two largest, but smaller deserts also include the incised and complex rocky outcrops of the American Southwest (actually several distinct deserts with unique biodiversities), the highlands of the Pamir alpine desert (where the greatest mountain ranges of the world meet in a jumbled pile of topography: the Himalaya, Hindu Kush, Karakorum, and Tian Shan), and the amazing Atacama Desert in Chile, which has many locations that have never recorded a drop of water from the sky since people started observing. To describe deserts is like describing islands—they are all unique in their own way, special places isolated climatically from their surroundings.

Wild Cotton

The "claws" of
the fruit provide black
fiber woven into beautiful
coiled baskets.

Devil's claw

Cholla

Sacred datura

Morning glory

Verbena

Desert lily

Barrel cactus

cal

na caltrop

Pincushion cactus

Desert Marigold

The specialized plants and animals in the desert make their living on the edges, finding precious hidden water and cool temperatures through uniquely evolved adaptations, skill, and timing.

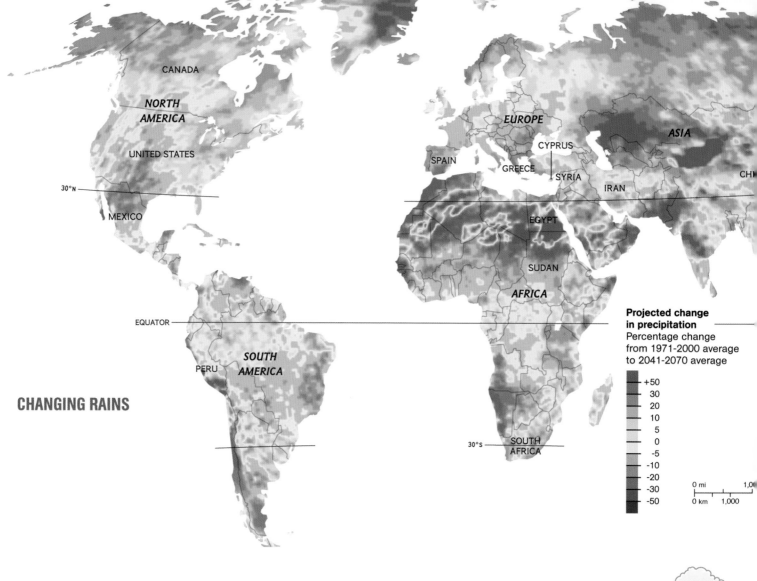

CHANGING RAINS

Projected change in precipitation
Percentage change from 1971-2000 average to 2041-2070 average

+50
30
20
10
5
0
-5
-10
-20
-30
-50

0 mi   1,0(
0 km   1,000

DROUGHT AND DELUGE

Warm air holds more moisture, carrying it away from dry areas (1) and toward wetter ones (2). Thus as global temperatures rise, dry areas will likely get drier and wet areas wetter. Seasonal extremes will likewise intensify, as moisture accumulated in the dry season is shed in downpours in cooler times, leading to seasonal floods in regions otherwise prone to drought.

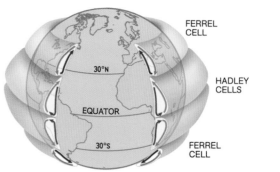

MOISTURE

WARM AIR

1

SPREADING DESERTS

Atmospheric warming is also predicted to affect rainfall by altering global air circulation. At present, warm air carried from the tropics by circulation loops called Hadley cells meets cool polar air carried by Ferrel cells in zones around 30° north and south, creating arid zones. As the planet warms, these zones are expected to expand and shift toward the Poles.

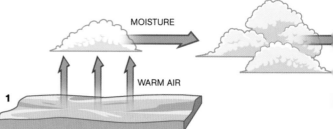

FERREL CELL

30°N

HADLEY CELLS

EQUATOR

30°S

FERREL CELL

Pres

SEAN MCNAUG
SOURCE: GEO

The world's systems direct water unequally, with deserts seeing the extreme deficits. But as the planet's atmosphere and oceans change, so does the distribution of that bounty. Many places will become drier, others wetter.

AUSTRALIA

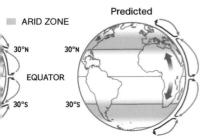

ARID ZONE
Predicted
30°N   30°N
EQUATOR
30°S   30°S

R. RITTER (MAP), HIRAM HENRIQUEZ (GRAPHICS), ALL NG STAFF
DYNAMICS LABORATORY, NOAA

The origins of deserts are, unsurprisingly, primarily climatic. This is apparent by looking at a globe; the major deserts of the world are clustered around 20 and 30 degrees north and south latitudes. These dry bands stretch around the world, across the southernmost parts of the United States and northern Mexico, through the entire continent of Africa, and across the scorching plains of India. In the south parts, the storied Kalahari Desert of Africa is connected to Australia in aridity, as is the driest desert on Earth, the aforementioned Atacama.

This global-scale pattern comes from general atmospheric circulation patterns. The sun warms the world the most at the equator, driving huge plumes of evaporated water skyward (recall the atmosphere mapping in that chapter). It cools and condenses into massive clouds; the ever-present rains of the tropics are a result of this incredible solar engine. The no-longer-wet air in the upper atmosphere continues to be pushed by the continued upwelling of warm, wet air and so must flow north or south. This dry air eventually cools and settles back down to the surface, creating near-permanent high-pressure systems in those subtropical latitudes, areas where the dry air is always flowing in, limiting the reach of any local storms (the Hadley "circulation cell").

There are exceptions, of course. Mountains and typhoons, for example, make southeastern Asia surprisingly wet. But generally the pattern holds—two bands of aridity stretching around the world, as a result of massive heat on the equator. This is also one of the reasons for the dry poles; the dry air, now on the surface, spills either north or south. As it flows along the surface it heats up, again beginning to rise as a result. As that air again rises, it partially drives the extremely wet latitudes around 50 and 60 degrees north and south, what is sometimes called the Ferrel cell. But what goes up must come down, and as you might expect, the dried-out, cooler air comes down either around the poles (completing the Polar cell, and creating the cold deserts of the Arctic and Antarctic) or back toward those warmer and wetter lower latitudes. The major deserts of the world are primarily the result of these large-scale atmospheric patterns of descending, dry air—patterns that are expanding poleward as the planet heats.

The other major driver of deserts, at a finer scale and particularly in midlatitudes, are mountains. The Ferrel cell as noted above is really a mixing of the more solar-driven Hadley and Polar cells and the air that descends on their edges tending to flow west (the "westerlies" on the northern edge) or east (the "trade winds" on the southern edge), all as a result of Earth's rotation (the Coriolis effect). As this air moves along the surface it hits mountains, rising and dropping its moisture as it cools. On the other side of the mountains the air descends, newly dry and heating up as it compresses due to higher pressure. This creates the localized deserts in southern Argentina, for example, behind the Andes Mountains, and is the second reason why the Atacama is so dry—it is situated between two mountain ranges, preventing moisture from both east and west. The third major reason is the presence of the cold Humboldt current just offshore, which creates a near-permanent temperature inversion, trapping warm air and further limiting ingress of moisture. In fact, the Atacama has been at this latitude and in this location for 150 million years and is likely the oldest desert on the planet, as well as the driest.

The stark categorization of these places into simple "deserts" belies their complexity. Like so many places, as much depends on the variability of the precipitation as on the actual amount. In all but the coldest places, life is found in astounding variety; only the continuous cold of central Antarctica can shut out life entirely. Life needs water, and frozen water is not particularly accessible. But wherever liquid water exists, however infrequent or temporary, life survives. The variability in deserts is created at a fine scale, both spatial and temporal. Cracks in rocks that collect water offer an ample supply of moisture in seemingly arid landscapes; rare thunderstorms allow plants to stock up on water.

As Edward Abbey, longtime curmudgeon (and he would consider "curmudgeon" a compliment) of the deserts of the American West realized in *Desert Solitaire*, "There is no shortage of water in the desert but exactly the right amount, a perfect ratio of water to rock, water to sand, insuring that wide free open, generous spacing among plants and animals, homes and towns and cities… There is no lack of water here unless you try to establish a city where no city should be." Deserts are islands of aridity on a blue planet; each desert is a perfect world unto itself.

Extreme dryness can require self-sacrifice of its specialized residents, however. *Rhanterium epapposum*, also known as *arfaj* in Arabic, is an example of how plants adapt to the variability and rhythms of desert life. A shrub native to the Arabian Peninsula and the national flower of Kuwait, *R. epapposum* is related to the common

WHEREVER LIQUID WATER EXISTS, HOWEVER INFREQUENT OR TEMPORARY, LIFE SURVIVES.

ABOVE RIGHT
Winter can bring occasional snow to deserts, dusting the rocks with white and providing important moisture that collects in crevices and filters through rocks all year long. Here, the Grand Canyon lies under a brief, white blanket in Arizona, USA.

RIGHT
Snow gracefully counterpoints the red rocks of the Grand Canyon rim, highlighting the layers and layers of desert sandstone, shale, granite, and other stones.

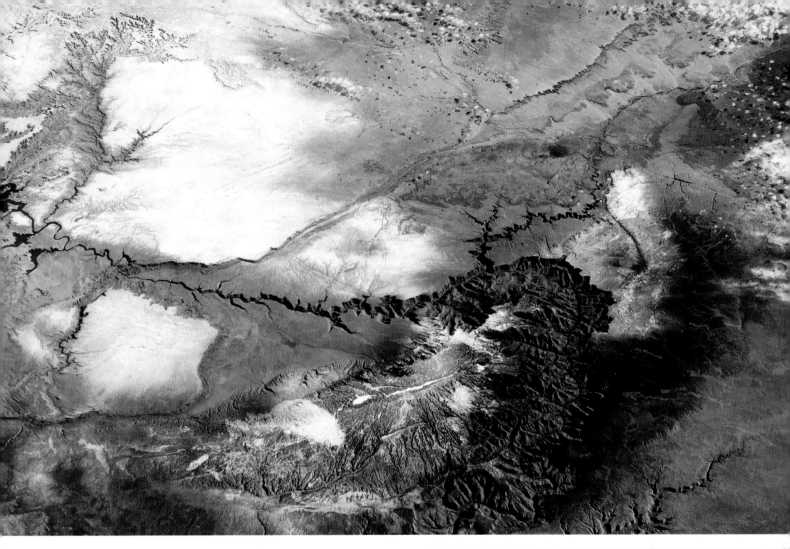

dandelion. In good years, the plant goes wild, sending shoots in all directions, rapidly producing thin leaves and twigs tipped with yellow flowers; it quickly becomes one of the most important forage plants for camels and sheep. But the good years give way to bad. In Dhahran, the eastern edge of the species range, a storm in 1997 dropped roughly 2.5 in. (7.6 cm) of rain, and another in 2002 dropped 17 in. (43 cm) in a day—but in the five years between those storms, rain was only recorded three times; less than 0.04 in. (1 mm) at a shot. In these extreme droughts, the plant cuts off its limbs to reduce water loss, forming a barrier of cork between the stem segments. Sacrificing growth and metabolism in all those hard-won branches means water loss is reduced to a bare minimum, as only a fraction of the canopy is left in contact with the roots. Willing to lose a part to save the whole, the flexible arfaj is one of the more common and successful organisms in its pocket of the world.

Each desert of the world is unique, a wondrous place where life is at its most inventive and the variability of plants and animals matches the unpredictable but precious moisture available. Few desert plants can compete if rains increase; other organisms will invade and thrive. And life can only tolerate so much variance; increased droughts can push organisms past their limits. As go the rains, so go the deserts.

TUNDRA: THE PAST AND PRESENT, ALL AT ONCE

As a desert is defined by low water, tundra is defined by low temperature (there is some overlap; tundras can be dry enough to qualify as deserts). Frigid winters and brief, cool summers are the rule. Climatically, they generally do not get above about 50°F (10°C) in any month, which, you may recall, is the rough thermal isotherm for trees at the global scale that Humboldt defined so many years ago. However, they always have at least one month or so in which average temperatures are above freezing, meaning liquid water and life.

The tundra biome in the North creates an almost continuous ring around the pole, the pan-Arctic boreal region, gracing the rim of Russia through the US, Canada, and Greenland. Then it takes a brief hop over the North Atlantic (don't forget to visit Iceland!), dips into Scandinavia and is back in Russia. Plant life defines the colder, poleward limit of the tundra; once life disappears due to a lack of liquid water, ice takes over. But the life is small. Trees are absent except in small pockets, unable to sustain their tissues far from the ground. Plants that thrive do so by staying low, below the snow surface in the winter to avoid desiccating winds and shredding ice and taking advantage of the solar heating of the ground in the summer.

Sir John Ross's extensive explorations of the Arctic spanned decades and were essentially unparalleled among other polar explorers in both scale and scope. His second expedition, which took a grueling four years, included the discovery of the magnetic North Pole by his nephew, James Clark Ross. That expedition, which was given up for dead by the public for two full years, was successful in large part because of John Ross's relatively peaceful and cooperative engagement with local Inuit cultures, who had been working the landscape for millennia. This map, an early compilation of the known geography (written in German) was contributed by John Ross to the US Library of Congress in 1855.

As late as the 18th century, the Arctic remained almost entirely uncharted. This 1711 map, by Henry Overton, illustrates the lack of knowledge. Even where islands are noted, such as the eastern Northwest Passage area (top/northern section of the map), the representations are crude and inaccurate.

And what a diversity of plants there are! More than 21,000 species of animals, fungi, and plants have been found in the Arctic north, and many circle the globe. In the South, there are patches of tundra on the Antarctic Peninsula, where the dominant plants are mosses, lichens, and liverworts, so-called "lower plants," without a vascular system. There are only two species of vascular plants on Antarctica proper: the yellow-flowered, pillow-shaped *Colobanthus quitensis* (Antarctic pearlwort) and plain-looking but clearly tough *Deschampsia antarctica* (Antarctic hairgrass). As a result, the tundra looks very different in the South than in the North. Sprinkled throughout the world there are also patches of tundra on high mountains, places where the growing season is too short and cool. In all tundra locations, temperatures are quite variable, swinging from -75°F (-60°C) in the winter to occasionally 86°F (30°C) in the summer—a huge range. But that cool upper average, considerably lower than the heatwaves that now consume the far North, constrains life to the small, the close to the ground, the fast flowering, and the long suffering.

Thawing permafrost

One unique aspect of tundra is permafrost. In many locations, the average annual temperatures are well below freezing, cold enough that the ground is underlain by permanently frozen soil. The upper soil layers thaw seasonally, part of the brief summer that makes life possible. This is known as the "active layer" and can be as shallow as 4 in. (10 cm) or as deep as 6.5 ft. (2 m). But below that is frozen dirt, the true permafrost, which does not get above 32°F (0°C) in summer. There is a common but mistaken conception of permafrost as a layer of ice underneath the soil. In actuality, permafrost is exactly what you'd get if you took a handful of soil from your local park and put it in the freezer—still dirt, just cold.

Take another imaginary journey, this time not shrinking down like Alice, but rather boring into the earth directly. Picture walking into a narrow tunnel, drilled straight into a hill in a permafrost zone. Above the tunnel entrance are shrubs, some grasses, perhaps a thin layer of moss. The door (there must be a door) is insulated and hooked up to a refrigeration unit, because the act of drilling into the permafrost exposes it to the wider atmosphere, which means it will thaw if not actively cooled. So you open the door, walk in, and see… what?

The tunnel you enter is cool but not cold, below freezing but not by much, perhaps about 25°F (-4°C). It is also quite dusty and dry, as the moisture gets pulled out of the air by the ever-present chill. As you proceed, you will occasionally (but not often) see pure ice descending from the ceiling, generally wedge shaped. These wedges form

Although small pockets of permafrost can be found in the high, mountainous regions of the world (such as the Tibetan Plateau and the Alps), the vast majority is found in the high latitudes of the northern hemisphere, shown here.

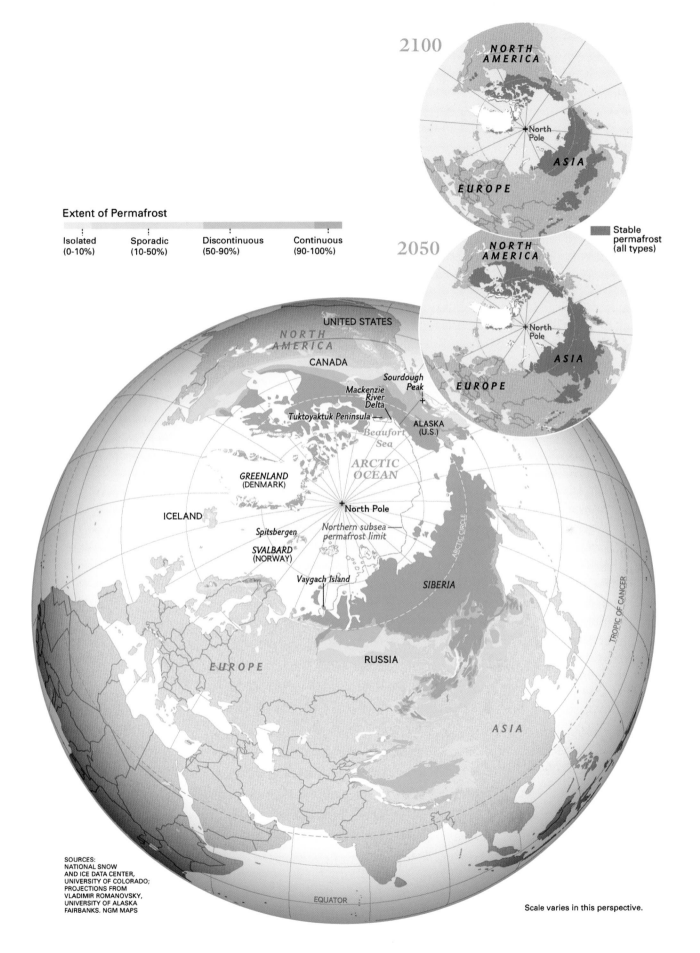

Extent of Permafrost

Isolated (0-10%)   Sporadic (10-50%)   Discontinuous (50-90%)   Continuous (90-100%)

2100

NORTH AMERICA

North Pole

ASIA

EUROPE

Stable permafrost (all types)

2050

NORTH AMERICA

North Pole

ASIA

EUROPE

UNITED STATES

NORTH AMERICA

CANADA

Sourdough Peak

Mackenzie River Delta

Tuktoyaktuk Peninsula

Beaufort Sea

ALASKA (U.S.)

ARCTIC OCEAN

GREENLAND (DENMARK)

North Pole

ICELAND

Spitsbergen

Northern subsea permafrost limit

SVALBARD (NORWAY)

Vaygach Island

SIBERIA

ARCTIC CIRCLE

EUROPE

RUSSIA

ASIA

TROPIC OF CANCER

SOURCES:
NATIONAL SNOW
AND ICE DATA CENTER,
UNIVERSITY OF COLORADO;
PROJECTIONS FROM
VLADIMIR ROMANOVSKY,
UNIVERSITY OF ALASKA
FAIRBANKS. NGM MAPS

EQUATOR

Scale varies in this perspective.

as the ground above your head contracts in the extremely cold winter, forming small cracks. During the brief summer season, the cracks fill with water, which then freezes in the winter. Ice expands when it freezes, about 9 percent by volume. The ice pushes the walls of the crack a bit larger. When it melts again the next summer, more water can enter because liquid water takes up less room. That water then freezes, making the wedge bigger. The growing ice wedge stops growing when the top stops thawing (perhaps when it is covered by vegetation), or if the wedge catastrophically melts and drains. This process can send the ice wedges vertically down into the soil—you will probably encounter a few in your tunnel, surrounded by that cool, dry dirt.

The most interesting thing you will notice is not the occasional ice wedge, though. It will be the small roots, branches, and occasional leaves poking out of the ceiling and walls. You may see little thin layers of charcoal lying horizontally in ice. Most shocking will be the bones. Bones sticking out of the wall, out of the ceiling, embedded in the floor. People have found whole mastodons. The amount of dead material in permafrost is amazing, dead life stacked upon dead life, frozen in place and looking like the day it died. Like a worm crawling through the soil, you are seeing the living world frozen in place from the bottom up. The deeper you go, the older the material; 10,000 years, then 20,000 years, and so on. Permafrost is full of this organic material, these roots, bones, and detritus, because it is just frozen soil. So the tundra is like two biomes in one. There is a thin, living surface layer full of shortgrasses and wildflowers, ponds (frozen soil doesn't drain, so standing water is common on the tundra), and seasonal migrations of birds and caribou that feast on the flush of plant life. But below that is a frigid realm, the world 10,000 years ago, literally frozen in place as if the clock stopped yesterday. That world is on ice, removed from our world by the simple trick of freezing. But it is coming back into play.

The amount of former life you will see in your tunnel is astounding. Its age is amazing. And it is all there, thanks to the cold conditions that stop decomposition. Things do not decay when frozen. What is the process of decay? Decomposition occurs via the same process that powers you and me: organisms eating carbon molecules (organic material), turning it into energy, and releasing the carbon as CO_2 or CH_4 (methane). But this is old carbon, a microbial-fossil fuel; it has just been put in the freezer for a few tens of thousands of years. What that means from a current climate perspective is a vast store of additional CO_2 and CH_4, which will be released as the frozen organic material thaws. We know where the bodies are buried, and when they show up, the climate will be upended dramatically.

MOST SHOCKING WILL BE THE BONES . . . PEOPLE HAVE FOUND WHOLE MASTODONS.

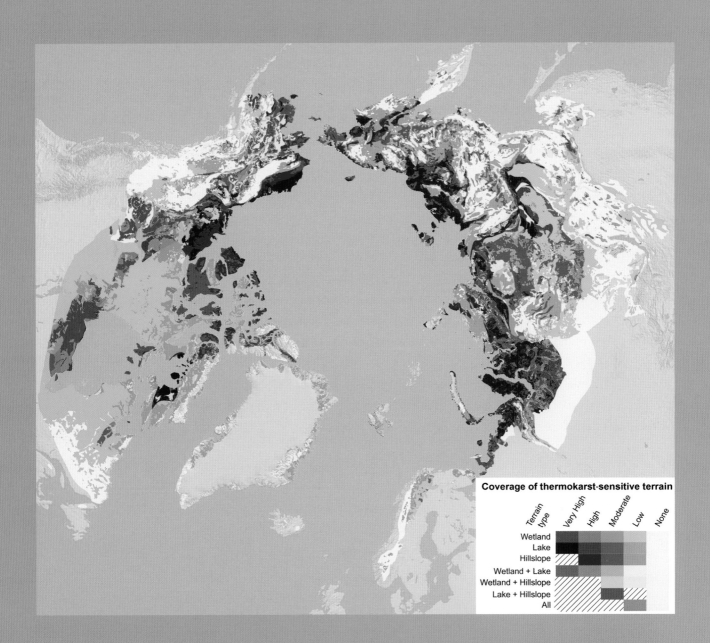

Coverage of thermokarst-sensitive terrain

Terrain type	Very High	High	Moderate	Low	None
Wetland					
Lake					
Hillslope					
Wetland + Lake					
Wetland + Hillslope					
Lake + Hillslope					
All					

When permafrost thaws, the ground erodes. The slumps and gullies that form are called "thermokarst," surface manifestations of the thawing permafrost below. Mapping the thermokarst landscapes offers a clue about how much of the landscape is susceptible to this rapid thaw. This map, built from satellite imagery and expert opinion, shows the dominant landforms crossed with the proportion of thermokarst features likely to be found or to form on the landscape.

The amount of carbon stored in permafrost and potentially released as CO_2 or CH_4 is enormous, about twice as much as all the carbon in the atmosphere already: approximately 1600 billion tons versus the current 850 billion tons. People often talk about "permafrost melting" but remember, it is mostly just frozen soil and organic material. It is more like placing a frozen chicken on your counter for a week. It will start to go bad, turn moldy and blackish, smell bad, and ultimately release that carbon as the microbes finish their feasting. It will decompose, and that carbon will enter the atmosphere. After all, it has to go somewhere.

Already, the Arctic region is a source of carbon to the atmosphere, rather than a sink—which means it is actively releasing carbon into the atmosphere, a positive feedback resulting from our warming of the atmosphere via carbon emissions. The most recent studies suggest carbon is being released faster than photosynthesis can take it in, mainly a result of decomposition over the winter months because in some places the ground is no longer frozen. Plant growth partially offsets the release in the summer, but overall it is not keeping pace with the annual carbon emissions from the thawing ground. Current estimates are that plant growth, and new plants expanding, will only offset about 20 percent of the carbon released by permafrost disappearance and subsequent decomposition.

THE RESILIENCE OF LIFE

But what of life? Many of our biomes have been defined by their variability rather than just their average conditions. The tundra is no different. Remember, the temperature swings that life must tolerate here are enormous, from -75°F (-60°C) in the winter to 86°F (30°C) or more in the summer. Species have evolved to tolerate a wide range of conditions; they are generalists, not specialists. Habitats within the tundra are incredibly diverse, from polar desert landscapes to wetlands. The tundra landscape is also the least fragmented of biomes, giving species room to track climate. We might expect them to be resilient: they are used to variability and well equipped to take advantage of whatever warmth they can get.

Experiments with plant life in tundra generally show a powerful response to warming. Mosses and lichens typically do poorly, likely because they lack a true root system and vasculature to move water around, meaning warming conditions cause water stress. More complex plants with roots and vasculature generally do better; shrubs grow taller, wildflowers increase in abundance. In an area where low temperature is the main constraint, a

TUNDRA REGIONS IN THE NORTH ARE WARMING FASTER THAN THE REST OF THE WORLD.

warming climate means many plants can take advantage. Growth explodes. Shrubs are encroaching on the grassy mid-tundra regions. Plants push poleward.

Another challenge is migratory life. The tundra is more than plants; it is a home for herds of caribou and innumerable birds. These migratory animals, especially birds, take the high-latitude bounty from the summer and head elsewhere—the boreal forest, the tropics, some as far as the opposite tundra (Arctic to Antarctic, and vice versa). There are approximately 200 species of birds that regularly breed in the Arctic tundra biome, 88 of which have the majority of their breeding range in the terrestrial wildflowers and sedges. Their prognosis seems to be a mixed bag, too, perhaps because of their differing life history strategies, perhaps because of our lack of knowledge. Land birds and wading species are declining, waterfowl are increasing. The cause is difficult to parse out; migratory birds pass through a range of landscapes before reaching the tundra—might the problem be there? It is difficult to say given the scale and scope of the question. Once again, our perception of the world is often the wrong scale of the problem, which can span hemispheres in the case of these migrating animals.

One potential culprit for the declines, which further illustrates the complexity of this landscape, is what is known as a phenological mismatch. Birds follow climatic and solar cues to know when to migrate. But the tundra regions in the North are warming faster than the rest of the world. Springs are arriving earlier and earlier there each year (at a rate of about a week per decade). A prospective breeding bird leaving its wintering site in Mexico plans on arriving just when the tundra's plants are at their most delicate—young, soft, and rich with nutrients. So the bird arrives at its usual time—but now that time is actually late. The plants are further along, larger, more resistant to browsing, a bit less nutritious. What's good for the plant is not good for the gander. This could be the reason for declines in some migratory species, or at least one of many contributing factors. Migration is one way an organism maintains a constant *experienced* climate in a world in which seasonal swings are incredibly large. Adjusting to a changing rate of those swings, especially when different parts of the planet are changing at different rates, is a hard nut to crack.

We might sum this up by saying the heart of the tundra biome is extreme temperatures and extreme temperature swings. Variability again is key, and timing that variability is a skill that organisms have mastered. When that timing is thrown off, however, all sorts of things happen—chaotic changes and rattling in the system. Some animals die off, some increase. Some plants spread, others retreat. And ever in the distance is that belowground rumbling of an ancient ecosystem reawakening and spewing its CO_2 waste into the atmosphere, speeding up the entire process.

CITIES

THE NEW ENVIRONMENT

Where do we put urban environments in this scheme of ours? Exclude it from the conversation, as cities are not part of the natural world? Embrace them as generated by animals (us) and thus just as natural as prairie dog colonies, beaver dams, and termite mounds? Whatever you think of the "naturalness" of cities, there is no denying that urban environments dominate large parts of the globe. The human footprint of development is interlaced with the natural world to such a degree that to not at least acknowledge the interface would be to miss quite a chunk of our landscape.

The human presence is so pervasive at this point that a recent proposal (as of 2021) to officially call this era the Anthropocene (a geological age during which human activities had an environmental impact on Earth) is winding its ways through the halls of the International Commission on Stratigraphy, the geological group that determines how the vast 4.5 billion years of Earth's history is broken into representative chunks. This is not a small deal. Part of the task is determining when the era officially starts—making it a reference point for all research going forward in geology, Earth sciences, and any other field which thinks long term. The date must be significant and, importantly, detectable around the world, so any researcher can find it in their excavations, wherever they are. (A leading proposal is currently the start of the Atomic Age, when nuclear bombs were being widely tested. This has left a layer of radioactive elements around the world that are highly distinctive. The human age arrived with a bang.)

HUMAN-CENTRIC CENTERS AS TRULY NOVEL ENVIRONMENTS

The new environments humans have created begin below the surface. Soil scientists, who spend their lives describing different types of earth, have designated new soil types found only in areas of long-term human occupation (in fact, there is a group with the impressive acronym of ICOMANTH, the International Committee on Anthropogenic Soils). These range widely. There are the "plaggic" and "haploplaggic" soils found under areas of long-term farming in northern Europe; the name is derived from *plaggen* agriculture, a practice likely invented in what is now the Netherlands, Germany, and Denmark, that spanned at least 3000 years. It involved harvesting sod in the summer, mixing that sod with manure or using it as bedding for farm animals, then reapplying it to the fields. There are also the aptly named "concretic" soils, thick layers of large concrete chunks common to construction sites or old, buried roads. Then there are the lovely sounding "pauciartifactic" soils, densely packed with discrete manmade debris in thick layers that form in landfills and dense urban areas. These complicated names describe the various ways in which our agricultural efforts, our discarded toys, or simply our waste gets packed and preserved, then finally melds into a whole new type of dirt beneath our feet.

We have also created unique biology. The London Underground (London's subterranean rapid transport system, also known as the Tube) and the substantial underground built areas of many northerly cities host a unique variety of mosquito. Unlike

> THE NEW ENVIRONMENTS HUMANS HAVE CREATED BEGIN BELOW THE SURFACE.

William Hubbard produced this first representation of New England around 1677. The small European colonial settlements are located on key rivers and bays, with the surrounding topography barely described.

most high-latitude mosquitoes, this one does not hibernate in winter, bites humans more than other organisms (such as birds), and cannot breed with aboveground species of mosquitoes. Some argue that this species evolved with the creation of London's famous subway, one of the first large, below-the-surface ecosystems in an otherwise cool environment. This enabled a common species (typical house mosquitoes, *Culex pipiens*) to go underground and switch from primarily biting birds to biting rats and, during the World War II London Blitz, to finding sustenance in humans. These new mosquitoes slowly became their own fully fledged, unique members of society as *Culex molestus*—now found around the world in urban areas with underground components. Others say that while they are interesting, the subway-dwelling pests are

THE URBAN HABITAT, FOR HUMANS AND THEIR ECOLOGICAL COMMUNITY

Cities are unique habitat for wildlife, semi-designed and wild. Species have carved out their own fine-scaled existence within the larger, engineered human habitat. Although perhaps a bit familiar, the diversity is still impressive when you consider the urban system as a whole—a maze of sharp edges, 3D complexity, and cracking cement in which many species specialize and thrive.

Peregrine falcon
(Falco peregrinus)

The peregrine falcon needs two things: high nesting spots and ample prey. In the wild, river gorges and cliffy habitats are ideal. The concrete canyons of cities provide both as well.

Rock dove
(Columba livia)

The wild rock dove (or common pigeon) nests in rock bands and has been domesticated for millennia; the transition to vertical concrete was straightforward.

Common ivy
(Hedera helix)

Graceful, clinging English (or common) ivy grows adventitious roots that thread into rough, vertical surfaces, giving the vine a height advantage over competitors.

London Underground mosquito
(Culex molestus)

Underground, there is little indication of winter. Many mosquitoes are now active year round, biting mammals more than birds—with a unique genetic signature to match.

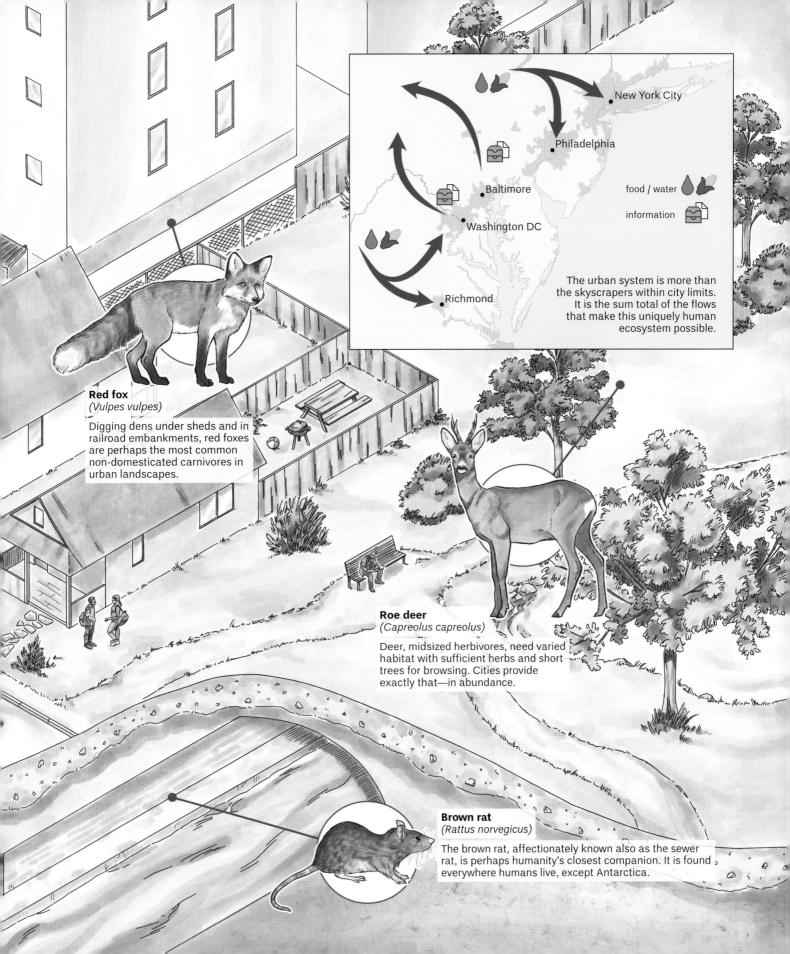

Red fox
(Vulpes vulpes)

Digging dens under sheds and in railroad embankments, red foxes are perhaps the most common non-domesticated carnivores in urban landscapes.

Roe deer
(Capreolus capreolus)

Deer, midsized herbivores, need varied habitat with sufficient herbs and short trees for browsing. Cities provide exactly that—in abundance.

Brown rat
(Rattus norvegicus)

The brown rat, affectionately known also as the sewer rat, is perhaps humanity's closest companion. It is found everywhere humans live, except Antarctica.

New York City

Philadelphia

Baltimore

Washington DC

Richmond

food / water

information

The urban system is more than the skyscrapers within city limits. It is the sum total of the flows that make this uniquely human ecosystem possible.

just a subspecies; somewhat different, but not warranting their own moniker (and so should be called *Culex pipiens molestus*). Others argue the development of the London Underground allowed southerly species of mosquitoes to colonize and evolve in the new urban landscape farther north. Regardless, it is clear there is a new, annoying kid on (under?) the block, and that a new urban realm created new habitat, and new possibilities, for life everywhere.

URBAN ECOLOGY

When people talk about urban environments in the language of ecology, they generally mean one of two things. The first is the *urban system* as an ecological system itself, complete and all encompassing. The second is the ecology *within* an urban setting. They are not the same thing, but they do overlap.

The "urban systems" definition places urban areas squarely in line with forests, coral reefs, and deserts—an energy-producing and energy-consuming collection of organisms, structures, and functions in relationships both visible and invisible. These links spiderweb out into the surrounding country; energy in the form of food streams into cities via superhighways, meaning the system itself is much larger than its city limits and concrete. This is no different than saying a lake system is larger than the lake; the lake system also includes the surrounding watershed, which contributes nutrients, food, and detritus to the fish populations within the lake. In this point of view, *energetics* define the system.

In the second definition, the ecology of a city is the life and interrelationships within it: the squirrels in a park, the web of tree canopies spreading down city streets, the polluted underbelly and the microorganisms that call it home. In this view, the city is the backdrop to a unique set of relationships created by that structure, a new collection of organisms that can tolerate or thrive in the urban pressure cooker.

Both definitions are correct; scale determines which is appropriate—city *as* an organism, or as *composed* of organisms? We will explore both. But first, we need to talk about what makes something urban.

How is a city environment unique? The question may be seem so obvious that it's not worth the time to ask, but there is value in describing what makes the urban jungle, well, urban. The previous sections of this book have spent a lot of time charting and describing the ways in which spatial variation in natural phenomena make the world diverse—how temperature variations around the world control the tree line, or how precipitation variability controls the geography of grasslands or glaciers.

And that variability occurs over vast scales, much larger than our limited personal perception. The same phenomena control life in cities: temperature, water, light, chemistry, energy. But in the case of cities, humans have some measure of control and design, and things are engineered to the human scale.

Light is an obvious example. Urban systems are disconnected from the dark, which is kept at bay by millions of sodium, neon, and blinkered LED lights. When those get disrupted, the whole environment changes. During the 1994 Northridge earthquake, large swaths of Los Angeles lost power and amazed residents flooded the Griffith Observatory with reports of strange, silvery clouds in the night sky. No doubt it was disconcerting to suddenly see clouds-that-weren't-clouds arcing over your city! Some unknown pollution? A spy installation? The reality was far more wondrous: it was the Milky Way, our own galaxy, a consistent fixture in the night sky since Earth started spinning. Billions of humans have seen it nightly throughout history, but not anymore. The starstruck inhabitants of LA were not alone in their confusing situation: 60 percent of those dwelling in Europe and 80 percent of us in North America cannot see our home galaxy. Eighty-three percent of the world's population and 99 percent of the populations in the US and Europe live with light-polluted skies.

This has serious implications beyond the aesthetic value of the night sky. An endless day, or some approximation thereof, alters circadian rhythms of organisms in a variety of ways—nocturnal animals have a harder time hunting, birds don't know when to nest, and metabolic disorders increase. "The city that never sleeps" has a poetic ring to it, but it also carries an unpleasant physical and psychological reality.

Water is another example. The perfection of concrete by the Romans is perhaps in the top 10 influential inventions of all time, but the art of mass paving was truly honed in the 20th century. As a result, cities are essentially large, impervious surfaces. Water ponds and pools, or runs off in sheets, causing water surpluses that include flooding in some places and deficits (like soil buried under concrete) in others.

Concrete soaks up heat as well, changing the thermal environment. Cities are slightly warmer than their surrounding countrysides, like small pockets of lower latitudes transplanted poleward around the world.

Species are different, too. Landscaping and horticulture have created grand plant experiments by bringing species from around the world into confined urban landscapes. Like a floral zoo, African tulips butt up against Hawaiian palms and Arctic grasses.

URBAN SYSTEMS ARE DISCONNECTED FROM THE DARK, WHICH IS KEPT AT BAY BY MILLIONS OF LIGHTS.

EXTENT OF DATA

Los Angeles

Miami

Mexico City

Caracas

Rio de Janeiro

Buenos Aires

FISHING
FLEETS

LIGHT POLLUTION
AROUND THE WORLD

The loss of dark night skies mirrors the rise of
urban ecosystems. Luminous patches glow on
a map of nighttime Earth created from satellite
and ground data on scattered light as of 1996–7.
The situation is even worse today. Based on
calculations, 83% of humanity lives under skies
polluted with light, and 20% can no longer
see the Milky Way. Least affected? The Central
African Republic.

EXTENT OF DATA

Tokyo

NATURAL
GAS FLARES

Mumbai
(Bombay)

Bangkok

CENTRAL
AFRICAN
REPUBLIC

Brighter night sky

Dimmer night sky

Area with sky glow

Dark sky areas

Johannesburg

0 mi   1,000

0 km   1,000

SEAN MCNAUGHTON,
M. BRODY DITTEMORE, AND
LISA R. RITTER, NG STAFF

MAP DATA COURTESY LIGHT
POLLUTION SCIENCE AND
TECHNOLOGY INSTITUTE

Cities are fascinating as they are, in many ways amalgamations of the historical natural and new urban ecosystems—created at the behest and design, intentional or not, of a single species. Like a concrete termite mound, we have created a habitat that is singularly unique in both its biotic and abiotic components. What makes this human habitat special is the combination of old and new, pieces of life collected from around the world slotting into niches created by new concoctions such as concrete, or adapting to new substances, from mosquitoes laying eggs in water collected in old tires to animal communities dealing with the introduction of synthetic estrogen in water supplies. These are massive experiments, even if unintentional and undirected.

FEEDING A CITY SYSTEM

If we think about urban systems the right way, they are like any other ecological system, producing energy, transforming energy, consuming energy, and excreting waste.

As a crude analogy, think of an urban area as a brain—an information-processing center, where decisions are made that generate actions over a much larger area. Like a brain, it is a highly interconnected, complex system that is extremely reliant on surrounding areas to route energy to power its functioning. Complex decisions with extensive spatial implications are made in cities; policies are envisioned and enacted, regulations designed and enforced. Like a brain, much of that organizational effort is oriented toward procuring energy and resources for the urban system itself. That energy flows in on arteries such as highways from food producing regions, the suburbs of the urban system reaching out like limbs into productive floodplains and fertile valleys. The energy is respired as CO_2 or is exported on the venous waste disposal systems to be buried, incinerated, recycled, or to meet some other fate. Urban systems are not just urban, they are the combination of the concrete and the croplands, the sum total of energy inputs and outputs that make unique "urban ecosystems" possible.

Early urban systems grew from the bounty of the landscapes that they integrated. The first cities of Eridu (around 5400 BCE) and Uruk (around 5000 BCE and one of the potential roots for the modern name "Iraq") were located where the arterial flow of major rivers consolidated the agricultural bounty of the Middle East—this surplus of raw energy, in the form of local strains of wheat, barley, peas, and refined goods like beer, allowed for the urban ecosystem to evolve. This immense landscape provided the energy, and the urban ecosystem provided the artists, priests,

THINK OF AN URBAN AREA AS A BRAIN, WHERE DECISIONS ARE MADE THAT GENERATE ACTIONS OVER A MUCH LARGER AREA.

tradesmen, and thriving bureaucracy to coordinate. But like any system, when energy inputs declined it suffered. Uruk, the centerpiece of the Sumerian empire (one of history's first) was located on the Euphrates River, which was utilized for irrigation, transport, drinking water, and sanitation. But precipitation gradually declined over the 2000 years of Uruk's power, sapping the city's strength. Continual irrigation resulted in the salinization of agricultural lands and lowered productivity; as water evaporated, any dissolved salts were left behind. Other political powers rose, with their urban centers in more productive settings and conquered productive territories. In a final coup de grâce, the river itself meandered off, leaving the ruins of Uruk isolated on the drying and desertifying plains of Iraq. Today, Uruk sits silently beside an extinct and long-abandoned channel of the Euphrates, presiding over a dead ecosystem from the past.

In fact, the world is scattered with the remains of cities starved of energy that diminished with shifting political winds or changing climates. Mohenjo-daro in Pakistan lost its central place in trade and agriculture as monsoonal patterns, and potentially the Indus River, shifted east; today it is a dusty plain. The 700 Dead Cities of Syria were abandoned when conquest shifted trade routes away, stripping them of their energy inputs. The classic Mayan civilization is thought to have collapsed during a period of extreme drought—a 50 to 70 percent reduction in annual rainfall—and is now a mere jungle-covered ruin. Starved of water and energy, the people scattered and the urban centers died.

But the self-organizing nature of human society persists, and urban systems that could feed those civilizations' voracious appetites on the bounty of their landscapes have survived. The city of Damascus, thought by some to be the longest continually inhabited urban area in the world, has survived invasion, religious upheaval, political catastrophes, and climate shifts by virtue of its geographic location between Europe, Asia, and Africa. This key intersection between continents means a continual supply of trade, which has proven resilient to external changes. Or consider Nanjing and Shanghai, in China. Founded on a wide bend with a natural port on the Yangtze River, Nanjing could integrate the economic output and marketable products of the immense upriver landscape of southern China. River trade was paramount in China, and Shanghai, downstream at the mouth of the Yangtze, could not wield the economic power of Nanjing, which was just upriver and had first crack at goods coming from the vast Chinese interior. Despite waxing and waning political clout (Nanjing was the capital of China and the largest city in the world in the 14th century), the energy flows never ceased, and Nanjing's regional superiority was unchallenged.

FOLLOWING
Major urban centers are increasing in number at an exponential pace, as people are pulled to common localities for work and opportunity, willingly or unwillingly. This graphic shows the number of cities in the world with populations of one million or more in 1800, 1900, 1950, and 2010.

RISE OF THE CITIES

Urban centers of more than a million people were rare as recently as the early 20th century. Today cities of more than ten million are not uncommon; there are 21 of them, almost

1800

3 CITIES OF ONE MILLION OR MORE

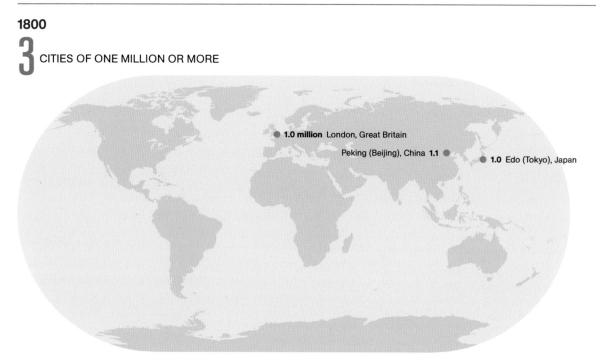

1.0 million London, Great Britain

Peking (Beijing), China **1.1**

1.0 Edo (Tokyo), Japan

CITY NAMES AND POPULATIONS REPRESENT URBAN AGGLOMERATIONS; THE LARGEST ARE LABELED.

1950

74 CITIES

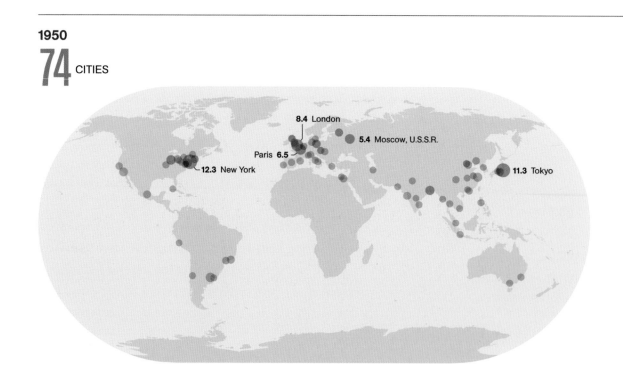

8.4 London

5.4 Moscow, U.S.S.R.

Paris **6.5**

12.3 New York

11.3 Tokyo

all in the developing regions of Asia, Africa, and Latin America, where population growth points to even bigger cities in the future. Metro areas have also overlapped to form huge urban networks; areas in West Africa, China, and northern India are each home to more than 50 million people.

1900

16 CITIES

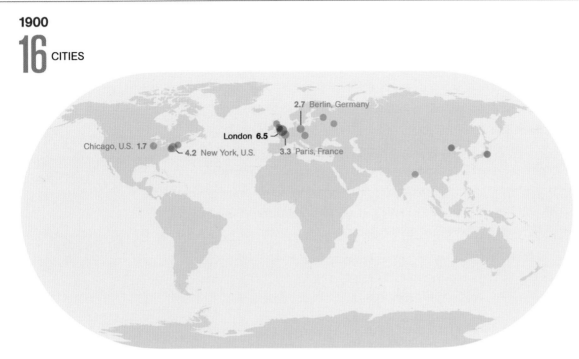

2.7 Berlin, Germany

London **6.5**

Chicago, U.S. **1.7**

4.2 New York, U.S.

3.3 Paris, France

● CITIES NEWLY ADDED TO EACH MAP ARE IN BLUE.

2010

442 CITIES

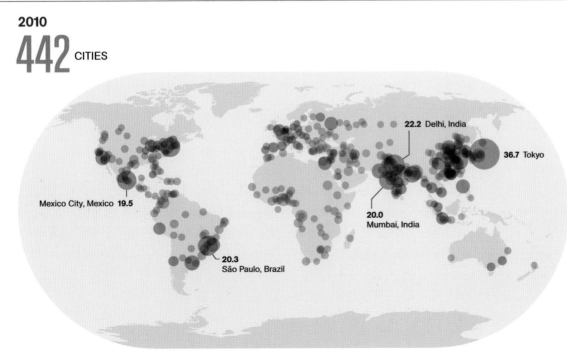

22.2 Delhi, India

36.7 Tokyo

Mexico City, Mexico **19.5**

20.0
Mumbai, India

20.3
São Paulo, Brazil

THE FIVE NATIONS WITH THE MOST CITIES OF ONE MILLION OR MORE: China **89** · India **46** · U.S. **42** · Brazil **21** · Mexico **12**

Only in recent years—as Shanghai's contributing energy landscape has begun to encompass the world via transoceanic trade—has that trade power eclipsed river-bound traffic, and coastal Shanghai has started to surpass her neighbor-rival.

One way to chart urban systems (meaning the whole system, defined by the energetics of inputs and outputs, farm to table) is surprisingly straightforward, a way that reminds us that we are, indeed, only animals living in a much larger world. This urban-system charting is done by mapping landscape productivity and human appropriations of that productivity. The productivity of any given spot on the planet can easily be defined in terms of carbon captured by the plants at that spot per unit time (for example, grams per year). Our bodies run on carbon, in particular glucose, a sugar molecule that contains six carbon atoms. Plants make glucose from sunlight, water, and carbon dioxide (and they run their bodies on glucose as well). Thus, carbon is the common currency of energy calculations: plants make it, we eat it.

Carbon productivity (sometimes called carbon fixation, because it takes carbon from the atmosphere and fixes it into stable, biological molecules) is measured as gross primary productivity (GPP), meaning the amount of carbon taken out of the atmosphere and converted into glucose by plants, or primary producers. Plants, like animals, have to expend some energy just to live, and so a more salient metric to the urban conversation is net primary productivity (NPP), or the amount of carbon fixed after accounting for that which is used by the plants to survive. Net primary productivity can be thought of as the "available" energy for other organisms to potentially utilize—bees taking nectar, worms eating fruit, or urban centers consolidating grain. Much like a farmer can live off the NPP of the fields around his or her house (in the form of corn for food, wood for structural materials, and so on), our cities mentioned earlier, from Eridu and Uruk to Shanghai, are all operating off appropriation of NPP from their energetic landscapes to the urban core. If available energy declines, urban systems must find new sources of energy or also decline. If energy increases, the surplus can translate into growth. Consider rethinking urban systems; not just as aggregations of people, but also as an emergent system built by concentrating carbon transfers (or things which can be traded for productivity, like money) into a single location. The people follow.

Our new conception of urban systems, then, extends far beyond the concrete and freeway-lined perimeter of traffic and suburbia. Tendrils of urban systems extend from New York City into the cornfields of Iowa, from London into the

IT IS INTERESTING TO TRACE HOW WE HAVE SOUGHT TO REFINE THE URBAN LANDSCAPE, TO TAILOR OUR HABITAT TO OUR DESIRES.

wheat-producing steppes of Ukraine, and from sunny Los Angeles into the cleared rainforests-turned-cattle ranches of Brazil.

One critical factor in this energetics view of the urban system is fossil fuels. Burning fossil fuels is simply appropriating NPP from a bygone era, productivity that was sequestered deep underground. The concept is exactly the same: taking carbon fixed by plants and using it for some purpose (in the case of fossil fuels, burning it for energy as heat or power). We could do the same with wood, it is just that oil, gas, coal and their relatives are highly compressed and therefore *much* more energetically dense. They are also quite stable, making them the perfect fuel. We are still utilizing NPP, just in a form that has not seen the light of day in many millions of years. The fossil fuel costs associated with producing and moving current non-fossil NPP from production zones into the urban core, and then moving the waste out, are difficult to account for—and what happens when that transport is lost? What if we run out of ancient NPP to power current NPP needs? If the ancient cities that populated Sumer and Pakistan (and basic math) are any indication, when energy ceases to flow into cities and an energy deficit occurs, the cities collapse and people move elsewhere. Like any system, built-up or otherwise, inputs must equal outputs. Currently, the inputs to urban systems are heavily reliant on the fossil-fueled transportation network. Just as when the Euphrates River wandered away from Uruk, depriving the city of upstream food and trade, when transportation via fossil fuels becomes more limited and expensive, or dries up, we should anticipate major changes.

DESIGNING YOUR OWN HABITAT

While thinking about cities from a systems perspective reveals how the tendrils of urbanity infiltrate landscapes and regions, it doesn't describe the ecology within the urban system. Drop down a scale and you will enter the environment that grasses and pigeons, rats and tulips, trees and squirrels experience. Cities are diverse, harboring a huge variety of plants from around the world and a plethora of tolerant animals and insects that have made peace with the urban jungle. This ecology is critically important to the well-being of the organisms residing there.

At its core, though, a city is a very human habitat, one designed for and maintained by the desires and imperatives of human society. It is interesting to trace how we, as a society, have sought to refine the urban landscape, to tailor our habitat to our desires. One illustrative facet of that tailoring is the evolution of the urban forest.

Cities developed on natural landscapes of course, building around and within natural features as the first towns took root. But as concrete and stone began to yield to the powerful engineering technologies of modern society, the scene quickly shifted from a living landscape to a functional but sterile conglomeration of buildings, streets, and forums. Trees and natural spaces were primarily afterthoughts in planning, either lucky legacies of the past found in empty lots, or ensconced in private courtyards for personal enjoyment.

Although city planning and open spaces date back millennia, the sprucing up of the modern city took off in the late 16th century, when the French democratized an Italian tradition of walled city gardens by taking down the walls and allowing the public to (occasionally) appreciate the display of life. Termed *allées*, these promenades were clearly an amenity—privately controlled, often by royalty, but occasionally publicly accessible. This was likely the beginning of the tree-lined avenue common in many cities today. The Dutch were similarly well known for having tree-lined canals around that time. But the overall aesthetic in urban systems until the 18th century was a functional one, dominated by stone, dirt, and industry.

This was also true in the early United States well into the 19th century, although a few cities passed laws protecting public trees within city limits (in 1656, Salem, Massachusetts, passed a law restricting wanton cutting of municipal trees, a surprising conservation statement from the community that would popularize witch trials only 40 years later). A good example of the evolving view of the urban forest is the town of Litchfield, Connecticut, one of the United States' early incorporated cities (1719) and peppered with history. In Litchfield's first years, the "only use for trees was to cut them down." Residents saw trees as a major impediment to urban development, blocking roads, breaking up

DC Shade Trees

By 1891, the shade trees of Washington DC were valued enough to be included in a collection of maps alongside property values, underground cables, and police stations. Significant trees are noted as numbers in red on the street map; numbers refer to species as listed in the upper right "Reference" section. For example, an American elm was found across the street and to the east of the White House (the "Executive Mansion"), denoted as #23. Elms were long favored as street trees for their rapid growth and shade, especially in the United States. American elm is now endangered due to Dutch Elm disease, introduced in 1928, only a few decades after this map was made.

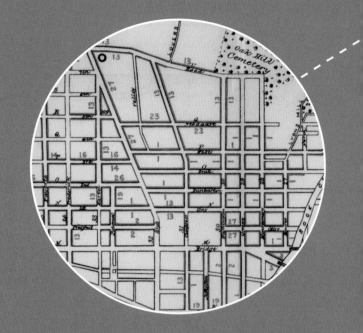

— Reference. —

No.	Botanical Name.	Common Name.	No.	Botanical Name.	Common Name.
41	Acer campestre	Turntable leaved maple	51	Quercus palustris	Pin oak
1	" dasycarpon	Silver maple	34	" roburfastigiata	Upright oak
13	" platanoides	Norway "	50	" rubra	Red "
14	" pseudoplatanus	Sycamore "	48	Robinia pseudacacia	Locust
15	" rubrum	Scarlet "	49	Salix Babylonica	Weeping willow
16	" saccharinum	Sugar "	29	" pentandra	Laurel-leafed willow
	" nigrum	" (black)	12	Salisburia adiantifolia	Ginko
36	Aesculus hippocastanum	Horsechestnut	3	Tilia americana	American linden
34	Cedrella sinensis		2	" europea	European "
9	Catalpa bignonioides	Catalpa	10	Taxodium distichum	Deciduous cypress
27	Fraxinus americana alba	White Ash	24	Ulmus americana	American elm
26	Gleditschia triacanthos	Honey locust	23	" alata	Winged "
7	" sinensis	Chinese "	39	" compestris	European field elm
	Gymnocladus canadensis	Kentucky coffee	25	" racemosa	Racemic flowered elm
31	Glyptostrobus sinensis	Chinese cypress	47	" montana	Scotch elm
5	Juglans nigra	Black Walnut	46	Acer campestre austriaca	Austrian maple
37	Koelreuteria paniculata		4	" dasycarpon Weirii	Weirs cut-leaf maple
11	Liquidambar styraciflua	Sweet gum	45	" platanoides Schwedleri	Schwedleri "
17	Liriodendron tulipifera	Tulip "	43	" Tartaricum	Tartarian "
6	Negundo fraxinifolium	Negundo	44	Betula alba	White birch
20	Populus angulata	Carolina poplar	42	" nigra	Black "
21	" alba	Cottonwood	52	Broussonetia papyrifera	Paper mulberry
32	" Bolleana	Athenian "	54	Celtis occidentalis	Nettle tree
22	" graeca	Turkeystan poplar	53	Cercis Canadensis.	Judas tree
30	" grandidentata	Large tooth leaved "	55	Diospyrus virginica	Persimmon
40	" nigra fastigiata	Lombardy "	8	Juglans regia	Walnut
19	" tremuloides	American aspen	58	" cinerea	Butternut
18	Phellodendron amurense		59	Laurus sassafras	Sassafras
8	Planera crenata	Zelkova	60	Maclura aurantiaca	Osage orange
35	Platanus occidentalis	Sycamore	61	Morus alba	White mulberry
1	" orientalis	Eastern plane	62	Paulonia imperialis	Paulonia
33	Quercus bicolor	Swamp white oak	63	Populus candicans	Grey poplar
			64	Tilia europea argentea – (Silver linden)	

Rosslyn

POTOMAC RIVER

ZOOLOGICAL PARK

NATIONAL PARK

SOLDIERS HOME GROUND

Tidal Reservoir

Eastern Branch

Washington Channel

NAVY YARD

WIDTH OF STREETS AND AVENUES
North
South
East
West

Avenues

Statistical Map No. 4.
Showing the lines of
SHADE TREES.
CITY OF WASHINGTON
Compiled by Capt. W. T. Rossell, U. S. Engr's,
Assistant to Engr. Comm'r.
To accompany the annual report of the Commissioners,
District of Columbia.

SCALE

walks, and obscuring from view the signs of businesses. Digging out stumps is not an easy task today, let alone then, without the advantage of heavy equipment. But in heavily forested landscapes, trees were woody barriers to progress. It is true that a few dendrophiles in Litchfield planted trees as memorials, but they were symbolic sites, not forests. In 1779, resident Oliver Wolcott Jr. (later the US Secretary of the Treasury) planted 13 sycamore trees to commemorate the 13 colonies—some seem to still be standing. But even that minimal effort was not universally appreciated. At the time of the plantings, one Litchfield elder is said to have exclaimed, "We have worked so hard in our day, and just finished getting the woods cleared off, and now they are bringing the trees back again!"

The true development of urban forests, rather than single memorial trees, did not become official in the region until the turn of the 20th century. The rapid urbanization of the late 1800s had been hard on urban life. Installation of electrical poles and wires took down many a sidewalk tree; widening avenues for traffic (horse and carriage, at that point in time) meant many shady trees had to go, and pavement was laid down everywhere. To counter this, progressive individuals and cities decided to make the urban ecosystem official, with the creation of "tree wardens," city officials that would oversee, manage, and preserve the treed life of the citizens. The term "urban forestry" was coined in 1894 in Massachusetts, when the first classes on caring for urban trees—as opposed to forestry in the wilds ("forest wardens" dated to 1886)—were offered at what is now the University of Massachusetts. The state followed with a law in 1896 and expanded it in 1899, saying all communities in the state must have a tree warden to look after this important municipal resource. Several other states followed suit shortly thereafter. The urban forest became official.

In the United States, one of the most stunning examples of the power of trees to infiltrate our cities, to inspire city officials and the tree wardens who protect them, and to better the urban environment is Washington DC.

By the late 1800s, the central part of the US capital city was brimming with shade trees, though the bounty did not extend much past the central core of the city. Two streets (New York and Pennsylvania Avenues) even had trees running down their central medians. This was an amazing number of growing things for an extremely dense, extremely busy urban setting. It set Washington DC apart, unique in the country. In 1911, William Solotaroff, secretary of the Shade Tree Commission in East Orange, New Jersey, wrote, "No cities in America possess such avenues of fine shade trees as the city of Washington. While visitors admire the fine public buildings, every one of them will admit that the chief attractions of Washington are the beautiful avenues of shade trees which make it one magnificent park." Unintentionally prescient about our current climate crisis, he went on to say, "From the viewpoint of health . . . trees help to purify the air by absorbing the carbonic acid (recall, that is simply CO_2 when dissolved in water) that is exhaled by man. . . . Trees also help to modify the temperatures of our streets. The normal heat of summer is still further intensified in cities by the reflection from the pavement and the buildings." While the city was built on a (literal) swamp, the intentional inclusion of an urban forest made the summers bearable and the springtime delightful. The urban environment is not only unique because of its composition; the climate of an urban area is also different from its surroundings.

> THE URBAN ENVIRONMENT IS NOT ONLY UNIQUE BECAUSE OF ITS COMPOSITION; THE CLIMATE OF AN URBAN AREA IS ALSO DIFFERENT FROM ITS SURROUNDINGS.

The stories of Washington DC's shade trees are many, but the tale of the city's most iconic trees, the cherries, goes back to Eliza Scidmore, the first female writer, photographer, and board member of the National Geographic Society. She had visited Japan and fallen in love with the delicately pink springtime flowers ("the most beautiful thing in the world"), a sight which had enchanted the Japanese for a thousand years. In 1885, Eliza requested that the US Army plant cherry trees along the Potomac. They never got back to her. So in 1909, she enlisted the help of Helen Taft, first lady at the time, and the plan blossomed. In 1911, the first 3000 trees were planted (alas, these were killed by disease; a second crop was planted in 1912 and was successful). The cherry trees now draw over 1.5 million visitors a year and contribute to the unique, anthropogenic biome that is the urban core of Washington DC.

A CITY AS A WORLD UNTO ITSELF: CITIES CREATE CLIMATE

This novel environment has some properties that emerge from our human-driven mixing of plant and stone. A brief return to Solotaroff's farsighted praise for the cooling power of trees illustrates this. Climate change, acres of concrete, and just plain hot days mean amplified heat within urban centers. Concrete absorbs heat readily; surface temperatures can be 50–90°F (27–50°C) higher than air temperatures. This adds up: cities of one million or more are typically 1.8–5.4°F (1–3°C) warmer during the day and a stunning 22°F (12°C) warmer during the night than surrounding lands. This makes for an island of heat, a bubble that is literally a different climate than the areas around it. Cities are truly their own new, and novel, environment.

The heat island effect results from the rapid heating of stone and concrete by solar radiation as a result of its low specific heat (the amount of energy needed to

New York City
TREE CANOPY

Pelham Bay and
Split Rock Golf Courses

Van Cortlandt
Park

Inwood Hill and
Fort Tryon Parks

New York
Botanical Garden

Hart Island

BRONX

City Island

Claremont
Park

Long Island Sound

Rikers

MANHATTAN

LaGuardia

Grand Central Pkwy and
Cross Island Pkwy Junction

Bryant Park

Ravenswood and
Queensbridge Houses

Queens Zoo

QUEENS

Haberman

Msgr. McGolrick
Park

Forest Park

Railroad Park

Ellis and
Liberty Islands

Prospect Park

Shooter's Island
(uninhabited)

New Brighton and
St. George

Green-Wood
Cemetery

Ditmas Park

JFK

BROOKLYN

Jamaica Bay

Freshkills
Park

STATEN ISLAND

Seagate

Breezy Point

raise the temperature of a given amount of concrete by a degree). Water, by constrast, requires about four times as much solar input to warm by an equivalent amount. The inclusion of plants, with their high water content, thus effectively cools a land-scape. There's more. Trees provide shade, which can reduce surface temperatures by 50 percent. And perhaps the biggest impact of the urban forest on a city's climate is that trees "sweat." When we *perspire*, water evaporates from the surface of our skin, taking heat with it. Trees *transpire*, which means they pump water out through their leaves via evaporation. They do this for photosynthetic reasons, and a bonus result is heat dissipation. So in a sense, trees are the sweat glands of a city.

To stay with Washington DC as an example, in summer 2018 temperatures soared to 102°F (39°C) in concrete-centric parts of the city, those dominated by highways, warehouses, and neighborhoods sans trees. But nearby—literally just blocks away, in neighborhoods with substantial street trees—temperatures were a more manageable 94°F (34°C) and in city parks the air was a pleasant 85°F (29°C). Eliza Scidmore's cherry trees, and the grassy National Mall, held temperatures to around 90°F (31°C). So, the urban environment is not solely at the mercy of external climate, cities create their own conditions as well. Although urban areas are in a real sense an intrusion on the historical landscape, carved out of former forests or estuaries, they are also their own biome in and of themselves. When you enter a city, think of it as visiting a new place, one that should have its own guidebook to the climate, flora, and fauna, just like a visit to the tallgrass prarie or a New Zealand forest. The urban landscape is human-created habitat, but all the more unique for it.

URBAN ENVIRONMENTS: A CONCLUSION

Sometimes, when major construction (such as highway road building) is going on, areas start to collect water as drainage is shifted and concrete creates impervious surfaces. Water trickles and runs into new topographies (such as curved highway off-ramps). Formerly dry areas, perhaps grassy or shrubby flats, start to get water-logged. Slowly, the deeper-rooted shrubs die off, drowned from the bottom up. Some of the grasses survive on little rises, while the low-lying areas begin to change. More spring wildflowers take advantage of the abnormal concentration of water. Different birds frequent the landscape—no longer the species that nest on dry ground, but migratory waterfowl. In my part of the world, cattails would spring up (they are

> TREES PROVIDE SHADE, WHICH CAN REDUCE SURFACE TEMPERATURES BY 50 PERCENT.

Modern methods for shade tree mapping include satellite imagery as a base, overlaid with machine learning (statistical) techniques to train a computer to pick out trees from the background of grasses, weeds, and other green things found in cities. These trees are as valuable a resource as any other facet of the urban environment. Here, darker green indicates higher densities of urban trees across the boroughs of New York.

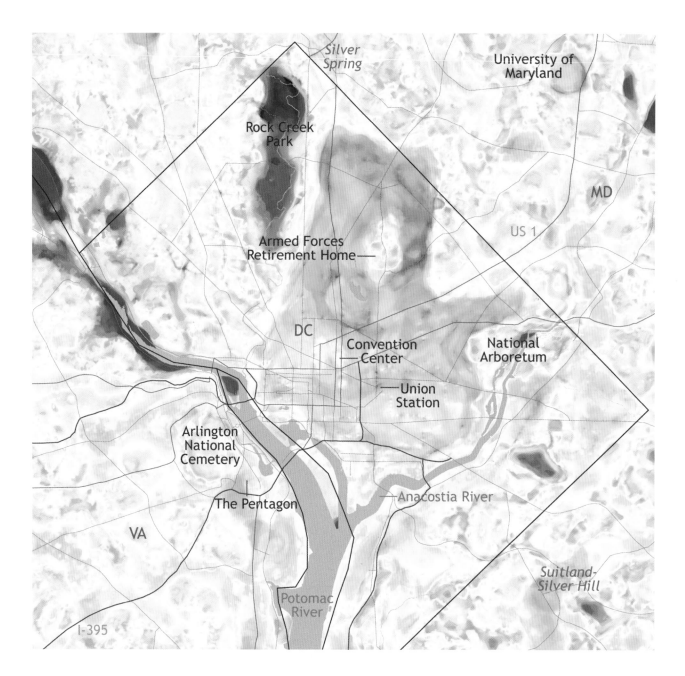

Afternoon air temperatures range more than 17°F (9°C) across a few places in Washington DC, going from pleasant to threatening. This difference is mediated by plant cover, clearly visible when the thermal environment is visualized at the citywide scale. The developed central core bakes while the parks and more leafy suburbs are much cooler, despite being only a short distance away.

Afternoon (3pm) UHI temperature (°F)

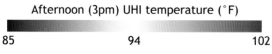

85   94   102

highly tolerant of pollution that inevitably runs off such roads). The transition would happen fairly quickly, perhaps over a decade, all because of a single change: water concentrated into areas where there was not water before.

Looking at the arc of human history, we see the same thing in urban systems. With the development of agriculture, humans began to trickle into the same location, to concentrate on the landscape. No longer roaming hunter-gatherers, we put down roots in a single place, and that place changed when we settled there. The first cities, small from today's perspective, were huge by historical standards. And like that water filling up a depression, the land in and around cities began to change. Unique, new soils developed over millennia. Some local species, or competitor species (like wolves) were eliminated. Other wild species survived in little pockets, persisting in abandoned lots or on the roofs of buildings. Some adapted and multiplied. New species moved in, opportunistically taking advantage of the new landscape or brought in by humans. The original urban landscapes were limited in their extent—novel environments embedded in a much larger, wilder world. Rivers, and later oceanic trade, extended that reach. The scale and scope of urbanity eventually extended far beyond its physical footprint by the acquisition of productivity from around the world, enabled by new transportation technologies.

We have built our own habitat, one as immune from the unpleasant variability in the natural world as we can make it. When it is winter in the northern hemisphere, we transport in our summer berries from Chile. When it is summer in the northern hemisphere, we export produce around the world. Immense tunnels transport water under mountains and between oceanic drainages to insulate us from low supplies in the summers. Air conditioning and heating make building interiors aseasonal; it's always tee-shirt weather regardless of the sleet outside. The urban habitat looks remarkably similar worldwide, a homogenization of conditions to which the natural world has responded. The world's most successful plants and animals are those that hitched their fate to humanity—the wheat, cows, rats, and pigeons of the world. Many complain about the problems of cities: the litter, noise, and casual violence against the natural world. I am guilty of that myself, at times. "Urbanism is the most advanced, concrete fulfillment of a nightmare," wrote Tom McDonough. But as a species, we nonetheless flock to cities for their opportunities, their resources, their unique conditions—in fact, more people now live in urban areas than outside them, globally. They have become the dominant human habitat, like any other animal and its chosen landscape.

WE HAVE BUILT OUR OWN HABITAT, ONE AS IMMUNE FROM THE UNPLEASANT VARIABILITY IN THE NATURAL WORLD AS WE CAN MAKE IT.

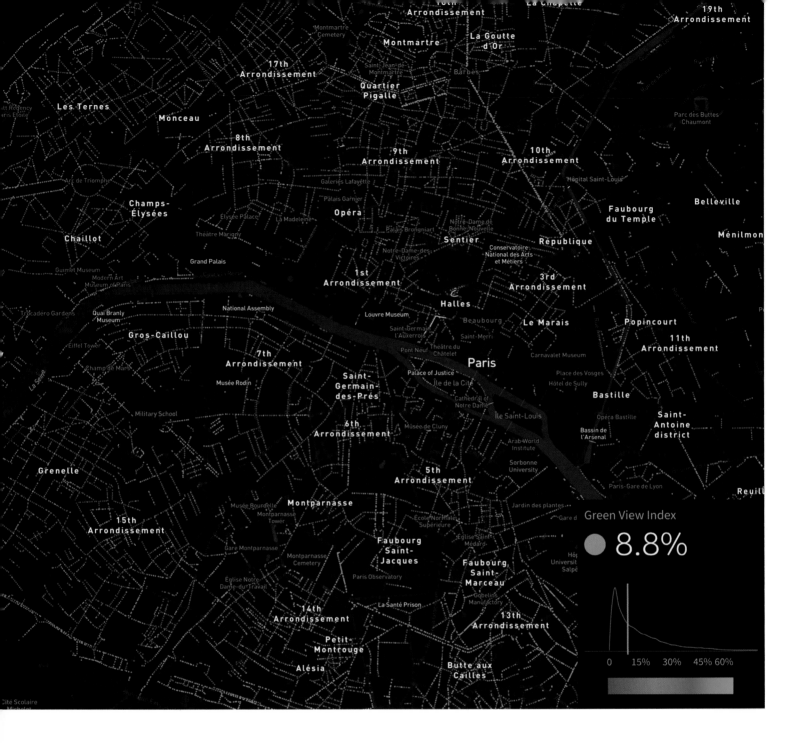

Paris and Seattle are both incredibly beautiful cities, but in differing ways.
The European City of Light is dominated by fine architecture and beautiful,
historic churches, while the Northwest tech hub revels in the lush, wet environment.
The "Green View" index, mapped here by the Treepedia project and expressed
in percentages, shows how present trees and vegetation are in these cities, as
perceived by a person walking down the street.

By using street-view imagery, each object covering the sky is classified as a tree canopy or building. If you stroll through a place with a lower index or percentage, you are more likely to be surrounded by buildings, stone, and soil. In contrast, walking through a location with a higher percentage predicts leafier coverings, shade, and a more natural experience.

LIFE

BIODIVERSITY

The land and seascapes discussed so far are the settings for life, the places where individuals play out their birth, life, and death by choice or necessity. We should talk about life. But it is hard to define a scale for discussing life, in all its varied forms and functions. Counting the microbial life that is our planet's most rich pool of species, there may be up to one trillion (1,000,000,000,000!) species on the planet. Each has its own story, its own evolutionary history, its own context. Each has its own color, smell, and habits. We can tell a few stories here, but beforehand, we should highlight some of the ways in which people talk about such diversity.

A fundamental challenge is the question of scale, because you cannot talk about life or biodiversity without defining the scale and scope of the conversation. For the broad chapters prior, we used coarse categories: the atmosphere, the oceans, the major land biomes. That was a necessity; if we want to tell interesting stories about each we cannot get into too much fine-grained detail. Therefore, we've focused on generalities, the broad patterns and basic physics that make the world tick. But if we wanted, those categories could be refined to finer and finer scales. Deserts that abut each other, like the Mojave and Sonoran, have very different flora and fauna. We can talk *all* deserts, or just *some* deserts, like the hot ones. Or just a single desert, like the Sonoran. But where does it end? We can talk about the biodiversity of Sonoran hillslopes. Or south-facing Sonoran hillslopes. What about the shady spots under boulders on those south-facing Sonoran hillslopes? Those are very different habitats from the sun-blasted open areas a few meters away. Alas, it appears our default of categorization to generalize is problematic, because it relies on lumping seemingly infinite variability into broad, bland categories of "deserts." And it appears this problem exists regardless of the spatial scale at which we choose to make our measurements or maps; we can always use more detail or less and categorize and describe different things. The more you zoom in, the more detail you observe.

Biodiversity thrives on this complexity and heterogeneity. Complexity (not just structural, but also in terms of life, resources, and habitat) generally means an increased number of options for life to specialize and thrive—and more interactions with other life, as well. As you look closer, you find more and more hidden in the nooks of the world, new life building on, and in, old life. The bumps in tree bark can harbor a vast array of tiny worms, beetles, and lichens in their shadows, worlds within worlds. When thinking about biodiversity, and where to find it, "a lifetime can be spent in a Magellanic voyage around the trunk of a single tree," E. O. Wilson poetically observed. It is true.

Very heterogeneous landscapes often have a wider diversity of organisms—coral reefs are prime examples. Though they represent less than one percent of total ocean, a full 25 percent of all marine species are found in their bounds. The complex geometry of the coral creates a variety of habitats for specialization: deep holes in the low-light zone, high-light environments, near-surface water, turbid sandy locations, protected inlets. Many millions of humans depend on this biodiversity for food, protection from the surf, tourism, and other uses. The Great Barrier Reef of Australia

"A LIFETIME CAN BE SPENT IN A MAGELLANIC VOYAGE AROUND THE TRUNK OF A SINGLE TREE." —E.O. WILSON

Golden scarab beetles, *Chrysina optima* (or *Plusiotis optima*), are also referred to as jewel scarabs. Found in Central America, their metallic sheen is highly distinctive and caused by the structural construction of their chitinous exterior cuticle.

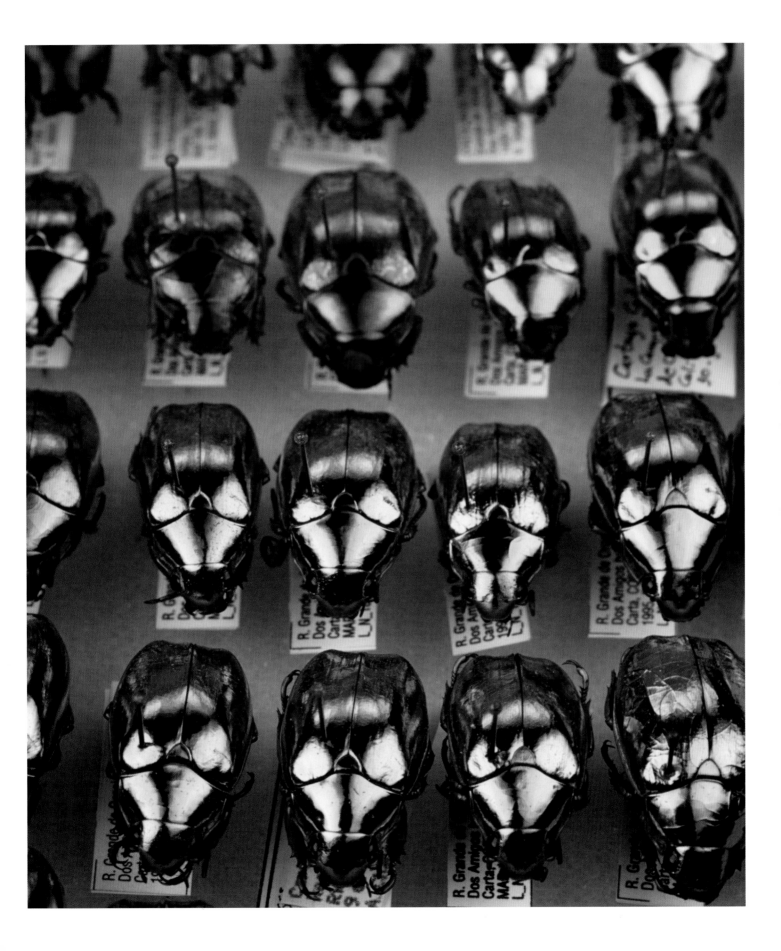

has 450 species of hard corals growing in fantastic shapes, from those formed like bowls to delicate fans extending many meters vertically in the water column. This underwater landscape supports 1625 species of fish, 3000 of molluscs, 630 echinoderms (think starfish or urchins), 215 bird species, 14 species of sea snakes, 6 (out of the world's 7) species of sea turtles, 30 species of dolphins and whales, 1300 species of crustaceans, 500 species of worms, 720 species of sea squirts, around 40 species of anemones, 150 soft coral species, and 133 species of sharks and rays, according to the Australian Great Barrier Reef Marine Park Authority. That is likely an undercount, given the complexity of the seascape.

For those of us not blessed to live near the reef, think of the moss underfoot. The complicated arms and fronds of the moss—with cavities that hold water and fringes that dry, the shadows and sunlight—are incredibly biodiverse. "One gram of moss from the forest floor, a piece about the size of a muffin, would harbour 150,000 protozoa, 132,000 tardigrades, 3000 springtails, 800 rotifers, 500 nematodes, 400 mites, and 200 fly larvae," notes Robin Wall Kimmerer in *Gathering Moss*. It's an astonishing diversity that most never see. Complexity builds on complexity, and those mosses are the richer for it. Just look closer, and you'll find more.

DESCRIBING DIVERSITY

Patterns of biodiversity are partially embedded in patterns of the nonlife environment, the forests crawling up mountainsides and the diverse willows and elk in narrow river drainages. But like complexity, life also builds on life and can create its own habitat, be it a coral reef or a city. Describing biodiversity *precisely* is an important step in understanding the emergent properties it creates and assessing how diversity changes with time and conditions (like climate change or urban development). But that is not a simple task. "Biodiversity" is a multifaceted property of a system, not a simple scale that moves up and down with the times.

Ask someone on the street what biodiversity means, and that person will likely say something to the effect of, "The number of species in a location" (if they have given much thought to the question at all). That is an aspect of biodiversity, certainly. But just a part. The number of species in a given location creates its species *richness*, a piece but not the whole of its diversity. Why the distinction? Because richness is only one facet, though perhaps one of the most important ones. Richness tells you how many pieces are present, but not how those pieces are arranged, or whether they are desirable or not, or how abundant they are. Like a puzzle, the richness or number of

Little islands of geology, serpentine rocks and the soils they create emerge from bedrock throughout the world. Shown in red in this map of California, these outcrops trace the mountain foothills, exposed by tectonic forces. Serpentine soils are toxic, with high concentrations of heavy metals like chromium and nickel. Few organisms can survive. As a result, unique plants evolved as serpentine specialists, tolerating the harsh landscape and thriving in their isolated, island-like patches. Overall, this means more biodiversity—a finer grain of variability and one that enriches the region.

Three Components of Climate, Measured by Color

Communicating the multifaceted nature of diversity is challenging. Biodiversity in particular responds to numerous things, including a variety of climate factors. This map attempts to communicate three components of climate simultaneously: temperature, precipitation, and seasonality (dubbed the Mediterranean Index here, meaning dry summers/wet winters). Lighter colors (approaching white) mean high values for all three; darker colors mean lower values. Areas with a high variety of colors, such as New Mexico, are hotspots of biodiversity in part because of the diversity in climates.

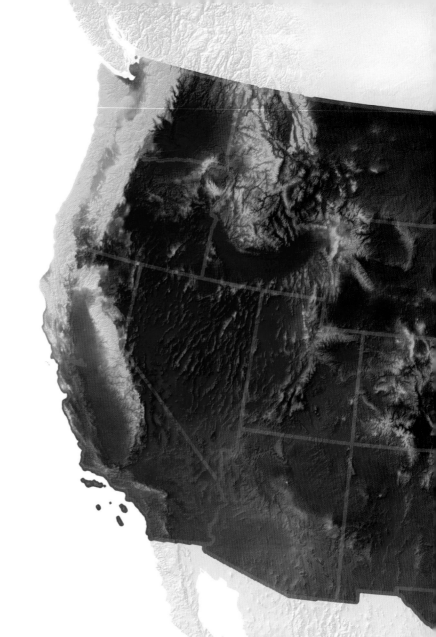

Increasing intensities of temperature, precipitation, and Mediterranean Index are shown as increasing intensities of red, green, and blue, respectively.

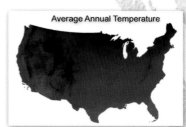

Average Annual Temperature

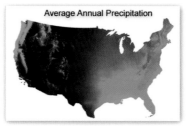

Average Annual Precipitation

Mediterranean Index

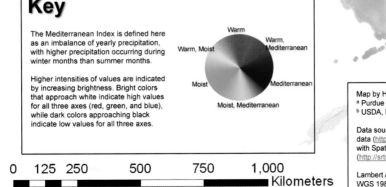

Key

The Mediterranean Index is defined here as an imbalance of yearly precipitation, with higher precipitation occurring during winter months than summer months.

Higher intensities of values are indicated by increasing brightness. Bright colors that approach white indicate high values for all three axes (red, green, and blue), while dark colors approaching black indicate low values for all three axes.

Warm

Warm, Moist

Warm, Mediterranean

Moist

Mediterranean

Moist, Mediterranean

Map by H.E. Winzeler [a], Z. Libohova [b], and P.R. Owens [a]
[a] Purdue University
[b] USDA, NRCS, National Soil Survey Center

Data sources: rainfall and temperature processed from PRISM data (http://www.prism.oregonstate.edu/), Hillshade processed with Spatial Analyst from SRTM elevation data (http://srtm.usgs.gov/index.php).

Lambert Azimuthal Equal Area Projection
WGS 1984.

0   125   250   500   750   1,000
Kilometers

pieces is a measure of the challenge in reconstructing the whole picture, but other things—the variety of shapes and colors and the overall beauty of the picture—are also important.

Consider the modern metropolis. Concrete stretching for kilometers, freeways and underpasses, skyscrapers and strip malls. Most would not consider these biodiversity hotspots. But look closer. Parks scattered here and there. Rooftop terraces. Meridians on the freeways. Plants from around the world, selected for their beautiful flower displays, adorning sidewalk restaurants. Little bamboo shoots sticking out of a dirt strip in front of a brownstone. If you consider species richness alone, cities can be fairly diverse. Unfortunately, there are not a lot of studies of how net species richness changes within a city.

One example we do have is from Adelaide, a city of 1.3 million on the south coast of Australia with a mild climate, ample rain, and gentle winters (though it does get the scorching heat waves that are increasingly common in Australia). Adelaide has records of species richness from 1836 onward (when Europeans arrived with their record-keeping habits) and has been resurveyed since. At Adelaide's founding, there were 1535 species recorded (all classed as native). After that, species numbers generally increased, to 1805 in 1900, 2036 in 1950, and 2047 in 2002. That is a substantial gain, and does not include garden plants that do not survive on their own, just wild plants (it does include garden escapees, horticultural species that can and did jump the fence and start wild populations). Almost all of that increase was in the plant realm; birds, reptiles, and amphibians all held steady (mammals declined, from 40 species to 29). These are from historical records, so likely have some error, as with the resurvey of Humboldt's Chimborazo climb. But the pattern seems to hold. Total species richness often increases in cities precisely because people introduce the plants and animals they know and love.

This should make it clear that species richness does not tell the whole biodiversity story, especially when it comes to the division between native and invasive/globally cosmopolitan species. Cities are not bastions of wild and exotic life; people explicitly go on vacations to see the wildlife they cannot see in their hometowns any longer. During that same time period in Adelaide, *native* plant species richness declined from 1136 species to 1047 and *native* mammals were halved from 40 to 20 species. The decline was more than offset by the introduction of new species, as we saw, but many of those new species weren't so "new." They were more along the lines of the common domestic cat (which decimates native bird populations) and dandelions. Sadly, the marvelous bettong, a rotund marsupial known as a "rat kangaroo" for

At its heart, biodiversity conservation is a political question. Decisions to set aside land, or work land differently, are made in particular locations, by particular people, and among competing interests. Maps must now communicate opportunities, accomplishments, and needs simultaneously. This graphic demonstrates one organization's successes and goals in New Mexico, an incredibly biodiverse state.

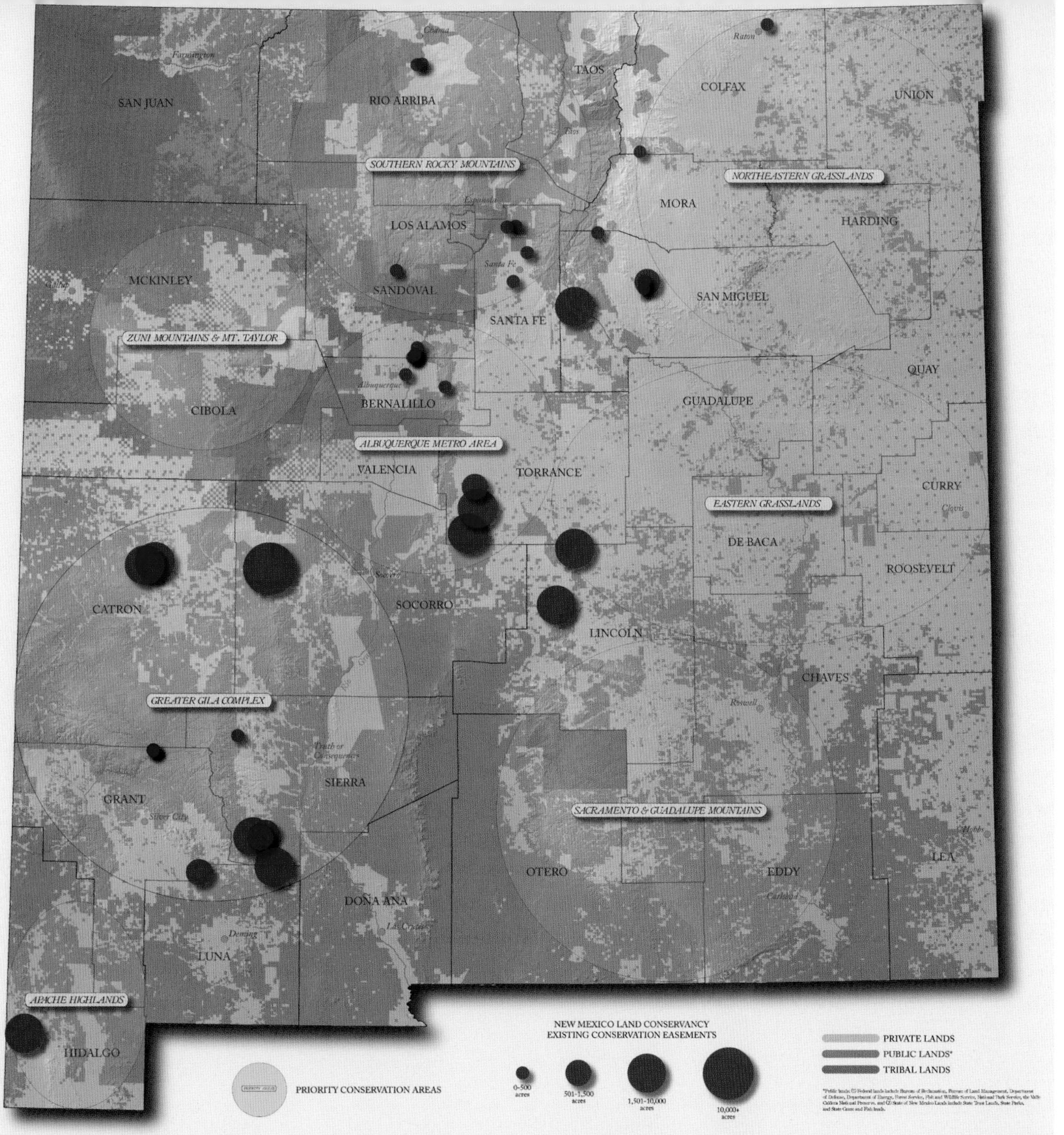

Map region labels: SAN JUAN · RIO ARRIBA · TAOS · COLFAX · UNION · SOUTHERN ROCKY MOUNTAINS · NORTHEASTERN GRASSLANDS · MORA · HARDING · LOS ALAMOS · SANDOVAL · SANTA FE · SAN MIGUEL · MCKINLEY · ZUNI MOUNTAINS & MT. TAYLOR · CIBOLA · BERNALILLO · QUAY · ALBUQUERQUE METRO AREA · GUADALUPE · VALENCIA · TORRANCE · CURRY · EASTERN GRASSLANDS · DE BACA · ROOSEVELT · CATRON · SOCORRO · LINCOLN · CHAVES · GREATER GILA COMPLEX · GRANT · SIERRA · SACRAMENTO & GUADALUPE MOUNTAINS · OTERO · EDDY · LEA · APACHE HIGHLANDS · DONA ANA · LUNA · HIDALGO

NEW MEXICO LAND CONSERVANCY
EXISTING CONSERVATION EASEMENTS

0–500 acres · 501–1,500 acres · 1,501–10,000 acres · 10,000+ acres

PRIORITY AREA — PRIORITY CONSERVATION AREAS

PRIVATE LANDS
PUBLIC LANDS*
TRIBAL LANDS

*Public lands: (1) Federal lands include Bureau of Reclamation, Bureau of Land Management, Department of Defense, Department of Energy, Forest Service, Fish and Wildlife Service, National Park Service, the Valles Caldera National Preserve, and (2) State of New Mexico Lands include State Trust Lands, State Parks, and State Game and Fish lands.

BERNALILLO COUNTY - 747,937 acres
Private 364,784 ac. — Private 49%
Public 155,479 ac. — Public 20%
Tribal 227,674 ac. — Tribal 30%

CATRON COUNTY - 4,434,524 acres
Private 1,134,497 ac. — 74%
Public 3,287,219 ac. — Tribal -0%
Tribal 12,808 ac. — Private 26%

CHAVES COUNTY - 3,888,051 acres
Private 1,935,343 ac. — Public 50%
Public 1,952,708 ac. — Private 50%
Tribal 0 ac.

CIBOLA COUNTY - 2,906,640 acres
Private 972,596 ac. — Public 35%
Public 1,029,587 ac. — Tribal 32%
Tribal 904,457 ac. — Private 33%

COLFAX COUNTY - 2,411,447 acres
Private 2,060,873 ac. — Public 15%
Public 350,481 ac. — -0%
Tribal 93 ac. — Private 85%

CURRY COUNTY - 900,696 acres
Private 839,002 ac. — Public 7%
Public 61,694 ac. — Private 93%
Tribal 0 ac.

DE BACA COUNTY - 1,493,648 acres
Private 1,208,446 ac. — Public 19%
Public 285,202 ac. — Private 81%
Tribal 0 ac.

DOÑA ANA COUNTY - 2,441,158 acres
Private 389,528 ac. — Public 16%
Public 2,051,630 ac. — Private 84%
Tribal 0 ac.

EDDY COUNTY - 2,686,393 acres
Private 597,903 ac. — Public 78%
Public 2,088,489 ac. — Private 22%
Tribal 0 ac.

GRANT COUNTY - 2,539,233 acres
Private 977,002 ac. — 62%
Public 1,562,230 ac. — Private 38%
Tribal 0 ac.

GUADALUPE COUNTY - 1,940,241 acres
Private 1,716,790 ac. — Public 12%
Public 223,451 ac. — Private 88%
Tribal 0 ac.

HARDING COUNTY - 1,360,605 acres
Private 932,573 ac. — Public 30%
Public 428,032 ac. — Private 70%
Tribal 0 ac.

HIDALGO COUNTY - 2,205,154 acres
Private 900,432 ac. — Public 59%
Public 1,304,722 ac. — Private 41%
Tribal 0 ac.

LEA COUNTY - 2,811,960 acres
Private 1,450,076 ac. — Public 48%
Public 1,361,884 ac. — Private 52%
Tribal 0 ac.

LINCOLN COUNTY - 3,092,046 acres
Private 1,696,435 ac. — Public 45%
Public 1,395,146 ac. — Tribal -0%
Tribal 465 ac. — Private 55%

LOS ALAMOS COUNTY - 69,924 acres
Private 3,991 ac. — Public 94%
Public 65,908 ac. — Tribal -0%
Tribal 25 ac. — Private 6%

LUNA COUNTY - 1,897,741 acres
Private 589,437 ac. — 69%
Public 1,308,274 ac. — Tribal -0%
Tribal 30 ac. — Private 31%

MCKINLEY COUNTY - 3,491,171 acres
Private 688,934 ac. — Public 19%
Public 649,133 ac. — Private 20%
Tribal 2,153,104 ac. — Tribal 61%

MORA COUNTY - 1,237,432 acres
Private 597,307 ac. — Public 15%
Public 185,409 ac. — Private 85%
Tribal 0 ac.

OTERO COUNTY - 4,241,338 acres
Private 470,045 ac. — Public 78%
Public 3,311,583 ac. — Tribal 11%
Tribal 459,710 ac. — Private 11%

QUAY COUNTY - 1,844,110 acres
Private 1,623,373 ac. — Public 12%
Public 220,737 ac. — Private 88%
Tribal 0 ac.

RIO ARRIBA COUNTY - 3,773,510 acres
Private 838,277 ac. — Public 57%
Public 2,144,971 ac. — Tribal 21%
Tribal 790,262 ac. — Private 22%

ROOSEVELT COUNTY - 1,570,861 acres
Private 1,353,816 ac. — Public 14%
Public 217,045 ac. — Private 86%
Tribal 0 ac.

SAN JUAN COUNTY - 3,544,368 acres
Private 972,307 ac. — Public 29%
Public 1,012,123 ac. — Private 6%
Tribal 2,294,938 ac. — Tribal 65%

SAN MIGUEL COUNTY - 3,030,621 acres
Private 2,446,875 ac. — Public 7%
Public 383,946 ac. — Private 93%
Tribal 0 ac.

SANDOVAL COUNTY - 2,377,207 acres
Private 483,989 ac. — Public 45%
Public 1,073,362 ac. — Private 21%
Tribal 819,856 ac. — Tribal 34%

SANTA FE COUNTY - 1,222,947 acres
Private 726,067 ac. — Public 33%
Public 404,047 ac. — Tribal 8%
Tribal 92,833 ac. — Private 59%

SIERRA COUNTY - 2,711,159 acres
Private 686,740 ac. — Public 75%
Public 2,024,419 ac. — Private 25%
Tribal 0 ac.

SOCORRO COUNTY - 4,255,180 acres
Private 1,294,801 ac. — Public 14%
Public 2,809,492 ac. — Tribal 0%
Tribal 150,887 ac. — Private 86%

TAOS COUNTY - 1,410,862 acres
Private 446,291 ac. — Public 60%
Public 847,662 ac. — Tribal 8%
Tribal 116,909 ac. — Private 32%

TORRANCE COUNTY - 2,141,333 acres
Private 1,638,412 ac. — Public 23%
Public 486,435 ac. — Private 77%
Tribal 16,486 ac.

UNION COUNTY - 2,451,392 acres
Private 1,945,400 ac. — Public 21%
Public 505,992 ac. — Private 79%
Tribal 0 ac.

VALENCIA COUNTY - 683,694 acres
Private 471,541 ac. — Public 11%
Public 73,374 ac. — Tribal 5%
Tribal 138,772 ac. — Private 69%

NEW MEXICO CONSERVATION INCENTIVES TIMELINE

The Land Protection History of The Nature Conservancy in New York

For over 56 years The Nature Conservancy(TNC) has protected land for conservation purposes. Since the first parcel of land was protected at the Mianus River Gorge in southern New York in 1954, TNC has led the conservation community in land protection. TNC pioneered new land preservation techniques such as the conservation easement and has fostered innovative partnerships. The science of conservation has changed dramatically since TNC was first founded. Currently, TNC implements a range of strategies to meet conservation goals, though land protection remains a critical piece of the conservation puzzle.

This map is a visual representation of the extraordinary past half century of TNC land protection efforts. We display the size of the individual transactions and their approximate location, clustered around the conservation project area. Some of these clusters of protection were conserved for wetlands, while others are focused on forest systems or freshwater. Regardless of the target, they all represent TNC's long commitment to preserving lands and waters for the health of biodiversity and people alike.

Timeline of Land Protection

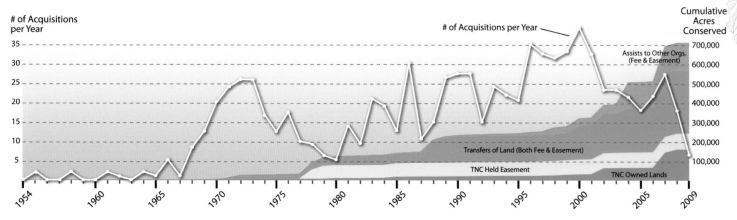

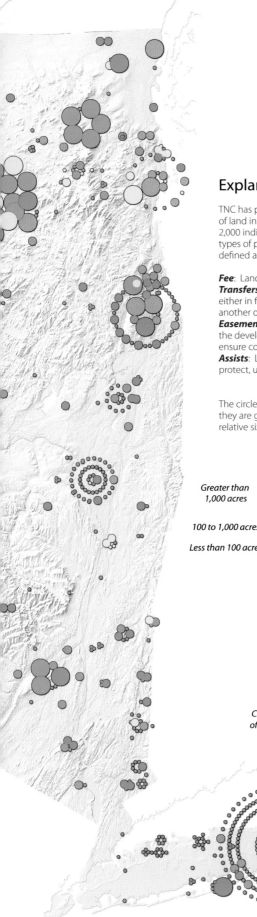

The Nature Conservancy

Protecting nature. Preserving life.™

Map designed by Brad Stratton

Explanation of Symbols

TNC has protected or helped to protect over 700,000 acres of land in New York with nearly 800 unique deals, and over 2,000 individual parcels since 1954. We employ several types of protection strategies. We identify four categories defined as follows:

Fee: Land that TNC purchased and still owns and manages.
Transfers: Land that TNC has purchased or had donated, either in fee or easement, and then transferred ownership to another organization, primarily NY state.
Easements: Land that TNC has purchased(or had donated) the development rights on and maintains monitoring to ensure conservation objectives are met.
Assists: Land that TNC has helped another organization protect, usually through financial loans or legal assistance.

The circles depicted on this map are not to scale. Rather, they are generalized cartographic representations of the relative size of the land deal.

	Fee	Transfers	Easements	Assists
Greater than 1,000 acres	●	●	○	●
100 to 1,000 acres	●	●	○	●
Less than 100 acres	●	●	○	●

25 Miles

Circular clusters are representations of of acquisitions, centered on the primary location of land protection

New York State has higher population pressures but no less need for conservation than New Mexico. This map, from The Nature Conservancy, clearly indicates that efforts are extensive—but biased toward certain portions of the state.

its look (large and rodent-like) and locomotion (kangaroo-like), was entirely lost in Adelaide, likely due to introduced predators such as the cats. But that is not reflected in simple richness numbers, the overall number of species.

Our tools for describing diversity clearly need some nuance. For instance, it might surprise you to learn that in a city in Europe of a given size, the number of bird species is generally comparable to the number of bird species found in a natural area of the same size. You might even cry "fowl" to that comparison, saying that you see all sorts of birds in the wild that you do not see in the city (which is typically dominated by pigeons, crows, and some highly adaptable raptors like the peregrine falcon). And your assertion would not be wrong. The issue is that outside the city, the variation in bird species you see occurs *across* different natural areas, from wild place to wild place. Cities as a whole tend to homogenize their landscapes. Nearly all urban spaces support broadly similar flora and fauna, at least in the key players: rats, cats, pigeons, dandelions (with some range of variation allowed for differences between, say, tropical and subarctic urban areas, of course; the monkeys in Kathmandu would be cold in Anchorage). This biotic homogenization is favored by all the features described in the urban section: the urban heat island which keeps temperatures consistent, a decrease in natural predators, and ample food resources—garbage, if you can eat it. In other words, urban environments create a trend toward consistency, not variability. The natural world has more net variety between places, more diversity across locations, more surprises around every corner. So our metrics of diversity could also include some sort of *spatial heterogeneity*, that same variability that is so important in all the other systems that we have discussed.

There's another angle to diversity that we haven't mentioned, one not captured by net numbers of species (richness) or variety from place to place (spatial heterogeneity). It is known as *evenness* or *dominance*. Neither richness nor spatial heterogeneity tells us about dominance. There is a big difference between a system of 3 species in which those species are distributed evenly (say, 33 percent of each) and one that is 98 percent species A, and only one percent species B and C. Evenness tells us if a single species has run away with the place, taken over, and the others are just hanging on by a thread. Even in natural ecosystems, some species are fairly dominant, and some are rare. In most ecosystems, though, the average species are common but not abundant. In human-dominated systems, our homogenization has led a few species to truly dominate—for instance, the pigeons in New York's Times Square. So just tracking

EVEN IN NATURAL ECOSYSTEMS, SOME SPECIES ARE FAIRLY DOMINANT, AND SOME ARE RARE.

the richness (number) of species through time may obscure important changes in evenness or dominance, as some species grow and swell their populations at the expense of others.

All this goes to illustrate that biodiversity is actually a complex, multifaceted characteristic of a system, and very much depends on what one is attempting to communicate. Species richness only? Non-native species culled? Spatial heterogeneity in the mix? Dominance included? Critical thinking—and precise communication—is required!

DIVERSITY ON THE MOVE: MIGRATION

The dynamism of the natural world in shaping diversity patterns (richness, spatial variability, and evenness or dominance) in both time and space are on display all year round in the world of avian migration. An immense diversity of birds move and swirl around flyways each year and migration highlights how that life responds to the variety of landscapes and seasons it encounters. Migration is also a fascinating look at a much finer scale, the scale of the individual, and how those individuals shape their experience of the natural world (its conditions and variability) intentionally, in ways that echo the human experience. Such movements also illustrate exactly why

At one time, a common explanation for swallow disappearance in the winter was hibernation. Olaus Magnus, a 16th-century Swedish writer, claimed that fishermen would occasionally find the birds quietly sleeping away the winter underwater in lakes and streams.

DE AVIBVS.

humans need charts and maps to visualize patterns that operate at a scale far removed from our limited eyes and ears.

Migration is a tricky puzzle. People no doubt observed birds appearing and disappearing seasonally, they just didn't know what was going on. In Job 30:26, the Bible mentions hawks moving south: "Does the hawk fly by your wisdom, and spread its wings toward the south?" Aristotle thought that swallows hibernated in holes, "denuded of their feathers," during winters in Greece (when they were nowhere to be found) and that redstarts and robins were the same bird, just changed in color in summer and winter. In actuality, as winter rolls in, robins are migrating into Greece, just as the redstarts are migrating south out of Greece. But it was a powerful idea at the time. By the Renaissance, people had generally abandoned the idea that birds morphed into different species, so the conjectures turned behavioral. In 1789, the noted naturalist Gilbert White still entertained the idea that swallows hibernated in winter in caves. One imaginative scholar, Charles Morton, rejected the hibernation hypothesis and speculated that swallows went to the moon. He reasoned that they could cover perhaps 125 miles an hour (a high estimate, but in zero gravity and with no air resistance, he thought it possible) and would be sustained by fat during the journey (which is generally true). And since the journey would take only 60 days (the distance to the moon was incorrect, though not terribly off; he calculated based on an estimate of 179,712 mi. or 289,219 km when in actuality the distance is 238,900 mi. or 384,473 km), the birds would be able to go back and forth each year. This was an interesting hypothesis which uses several real insights, like birds using fat reserves from summer to power their migration, and one very bad assumption ("if nobody knows where they go on Earth, they must leave Earth") to come to an imaginative conclusion about migration.

In fact, out of the approximately 10,000 bird species around the world, only one is known to actually hibernate: the common poorwill, *Phalaenoptilus nuttallii*, native of western North America. It is a small, nocturnal bird, about 7 in. (18 cm) in length, with a wingspan of nearly 12 in. (30 cm). It goes into a state of torpor, or extremely reduced metabolic activity, to conserve energy for several months each winter. The common poorwill seeks out rocky habitats to do this, places where extreme temperatures in the winter are minimized. Other birds just leave rather than endure winter cold.

What seems simple is difficult to explain without a global perspective. Today, of course, we know that birds migrate, but understanding the evolutionary origins of bird migration is still a challenge. Regardless, the ecological reasons are

OUT OF 10,000 BIRD SPECIES . . . ONLY ONE IS KNOWN TO ACTUALLY HIBERNATE.

Not all migrations are equal in length, but all require specific corridors and routes that avoid topographical barriers and provide both food and shelter en route.

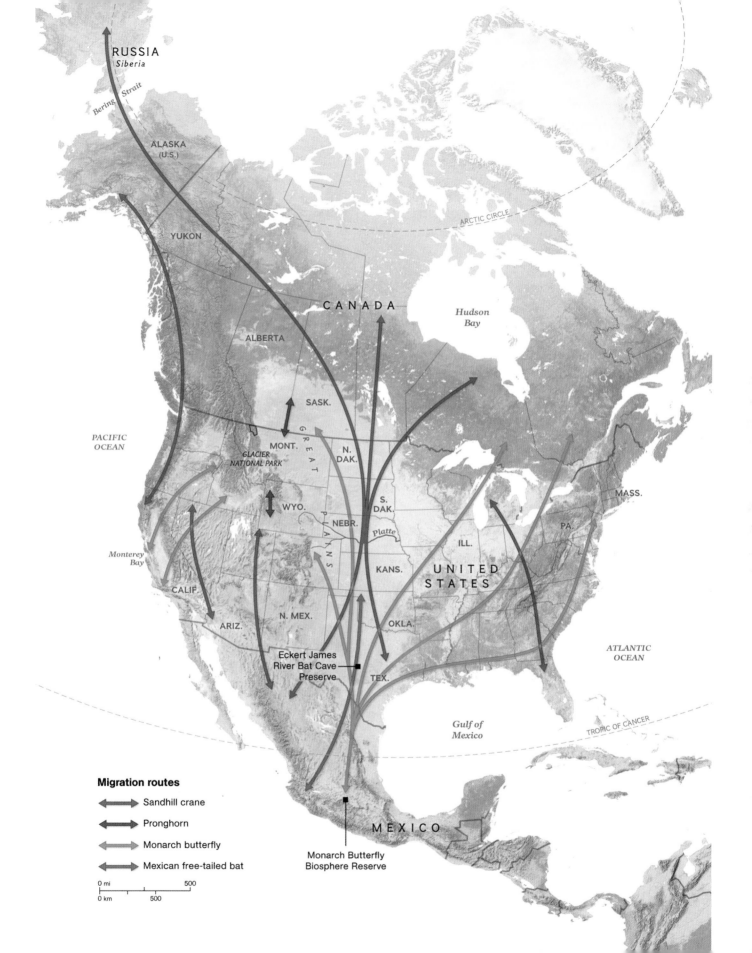

RUSSIA
Siberia

Bering Strait

ALASKA
(U.S.)

YUKON

ARCTIC CIRCLE

CANADA

Hudson Bay

ALBERTA

SASK.

PACIFIC OCEAN

MONT.

GLACIER NATIONAL PARK

N. DAK.

GREAT

WYO.

PLAINS

NEBR.

S. DAK.

Platte

ILL.

MASS.

PA.

Monterey Bay

CALIF.

KANS.

UNITED STATES

N. MEX.

ARIZ.

OKLA.

ATLANTIC OCEAN

Eckert James River Bat Cave Preserve

TEX.

Gulf of Mexico

TROPIC OF CANCER

Migration routes

Sandhill crane

Pronghorn

Monarch butterfly

Mexican free-tailed bat

MEXICO

Monarch Butterfly Biosphere Reserve

0 mi   500

0 km   500

FOLLOWING THE SUN

Many birds are highly mobile, traveling the world chasing daylight for food and breeding in their annual migrations, which can take them from the far North to the far South and back. Others endure, staying in singular home ranges their entire lives.

Mottled duck
(Anas fulvigula)

Year round: coast of the Gulf of Mexico

Common nighthawk
(Chordeiles minor)

Breeding range: North America
Winter range: southern South America (primarily Argentina)

American golden plover
(Pluvialis dominica)

Breeding range: Arctic tundra of Alaska and Canada
Winter range: central and southern South America

Canadian goose
(Branta canadensis)

Breeding range: Canada, Alaska
Winter range: southern U.S. to northern Mexico
Year round: northern and central U.S.

Little blue heron
(Egretta caerulea)

Breeding range: internal southeast U.S.
Winter range: Central and South America, Caribbean

Common poorwill
(Phalaenoptilus nuttallii)

Year round: western U.S. to northern Mexico

Piping plover
(Charadrius melodus)

Breeding range: Atlantic Coast and Great Lakes, northern Great Plains
Winter range: Gulf of Mexico, southern U.S. Atlantic coast, Caribbean

Western sandpiper
(Calidris mauri)

Breeding range: northwestern Alaska
Winter range: Pacific and Atlantic coasts of North and South America, Gulf of Mexico

straightforward enough: the costs of flying vast distances must be less than the costs of staying in place. It is a way to avoid detrimental variability in the natural world, to "create" their own lived habitat, by movement. Migratory birds that nest in northern Canada can, via migration, never experience winter. The American golden plover (*Pluvialis dominica*) moves from the Arctic to Patagonia each year, following the warm summer sun. The reason is clear: there are more resources available in the summer than in winter, regardless of where those summers are. There are also tropical migrants, which have the same logic, following or avoiding seasonal precipitation patterns. The birds are choosing their habitat, one that reduces unwanted variability or maximizes resources needed at a particular time, such as nesting or feeding. The Canadian Goose (*Branta canadensis*) is an excellent example. It moves north, approximately following the 35°F (1.6 °C) mean daily temperature, a strategy which means it generally finds open water and fresh-sprouting grasses. This is a slow journey, of course. Many other species choose to wait until spring is accomplished at their endpoint, and then migrate much faster (such as 130 miles a day, the reported rate for the gray-cheeked thrush, *Catharus minimus*). But the end result is the same, optimizing available resources by following the most productive climates and locations or by targeting areas that match specific needs.

This ability to "create" a continuous optimal habitat by movement is reminiscent of the human desire to build habitats uniquely suited to human needs or the ability of many species to use behaviors to avoid variability that is detrimental, like a bear hibernating through winter or an annual grass seeding and dying before a killing frost. The systems of the natural world can be understood by their patterns of variability, and life is adept at finding ways to optimize its use of those patterns.

LIFE AND CLIMATE CHANGE

But patterns of the world are being upended. Throughout this book are illustrations of how those patterns have been used for centuries by both human and nonhuman diversity to define the rhythms and seasons of a lifetime—seasonal patterns of fish, the weather, the rains in deserts. Variability can be useful, or not, but generally it is true that unpredictability is hard on life. Many plants carefully match the growth of their spring leaves to increasing day lengths, a tried-and-true way to time the growth of sensitive bud tissue with the beginning of spring. But with the destabilization of the climate, freezing events are occurring much less predictably than in the past, including late-season frosts that kill growing tissue. Other plants respond to warming cues, but midwinter warm spells are similarly becoming more frequent, tempting plants into spring growth long before the threat of a typical frost is past. Birds are migrating according to their age-old relationship with the spring, which now has them arriving at their nesting grounds far too late—the plants there have been growing for weeks due to polar amplification driving unequal warming around the world. Droughts are more intense, but so is the subsequent rainfall, resulting in mudslides and erosion. It is, of course, impossible to address even a fraction of the ways in which biodiversity is responding and altering with climate change, much less the myriad ways in which it can be charted. The response of life to a changing climate is perhaps the most active area of ecological research to date.

We can make an attempt at a few generalities as a way to tie together the patterns discussed here. To begin, we can talk about where climate change is first causing unprecedented impact, which tells us a bit about differences in vulnerabilities. After that, we can talk options.

Not all parts of the world are changing in the same way. On a broad scale, the whole planet is warming and precipitation is more variable; some places up, some places down. The world is heating fastest at the poles, and slower at the equator—which may tempt you into thinking that life at high latitudes has a worse prognosis. It tempts others into saying climate change is not a big deal—a few degrees warmer on average is nothing compared to normal summer-winter swings. It is true that temperatures at higher latitudes are much further above average than tropical latitudes. But it is not that simple. Tropical latitudes have much less variability in the temperatures they normally experience, a lower natural range. In the Arctic, 5.4°F (3°C) of mean warming is relatively little compared to the 135°F (75°C) swings common from winter to summer and the substantial year-to-year variability. But in the tropics, that same 5.4°F (3°C) may usher in a wholly unprecedented suite of conditions. The tropics are stable, temperature-wise, throughout the year; species may very well have *never*, in any season, experienced those new conditions. And the ability of life to adapt is closely related to the amount of time available between now and the unprecedented future climates they need to adapt to. It is important to remember variability, and figure out at what point climate change will enter uncharted territory, when and where even the coldest years will be warmer than the warmest year today.

Sometime between the years 2047 and 2069, Earth's land masses will, on average, be outside that historical range of variability in terms of temperatures; in other

THE ABILITY OF LIFE TO ADAPT IS CLOSELY RELATED TO THE AMOUNT OF TIME BETWEEN NOW AND THE FUTURE CLIMATES WE ARE CREATING.

words, in true untested waters. The oceans have slightly longer, until between 2051 and 2072. But that is the average, and if we have learned anything it is that averages mask the heterogeneous reality of the natural world; thinking about averages is wrong in both scale and scope. The low annual variation in the tropics means that even a small temperature increase can push them out of their historical bounds—the conditions to which life there is so exquisitely adapted. Tropical latitudes will pass entirely into the new thermal world 15 years earlier than more temperate latitudes. Some places, like New Guinea, are passing that unpleasant milestone—in which the coldest years are warmer than the historical warmest—in 2021. Jakarta, home to 9.6 million people, will enter a truly new climate world in 2029. Meanwhile, Anchorage, Alaska, where climate is warming twice as fast, will not reach wholly unheard-of temperatures until 2071.

Unfortunately, most of the world's threatened biodiverse communities are located in the tropics and on average, those hotspots will pass into truly new climates about a decade sooner than the global average. Coral reefs in the tropics—reefs that are acutely sensitive to temperature changes and often respond by bleaching and mass death—will hit unprecedented temperatures between 2034 and 2050. Mangroves and the incredibly diverse tropical-subtropical seagrass beds are on similar timelines. Hotspots for terrestrial mammals cross the threshold around 2037, give or take a few years depending on latitude. The only group of organisms slightly better off than the average are marine birds, because their hotspots are located at high latitudes, with a higher inherent range of variability. But even those will enter a new temperature regime by 2055 or so. This analysis of variability and climate change sets up a triple challenge for global biodiversity. The climate will truly enter a novel space faster in the tropics than anywhere else; life in those areas is precisely tuned to a low-variability climate, both intra- and inter-annually; and biodiversity is highest in exactly these regions.

OPTIONS: MIGRATE, ADAPT, OR DIE

Life only has stark choices when it comes to a climate that is leaving them by the wayside: migrate, adapt, or die.

Migration is possible if species can keep pace with the ranges of historical climate variability to which they are pre-adapted. So how fast is that? We can measure the velocity of climate change as the distance needed to travel to stay in a specific temperature range (or more commonly at some temperature average, since that is only a single variable to worry about in the calculations). It becomes a matter of tracing lines between a temperature at some location now and the nearest place you would find that temperature at some specified time in the future, figuring out the distance, and dividing by the time difference. In some ways that is an oversimplification, but it is a useful starting point for discussion.

For example, if your average summer temperature is 68°F (20°C), but 10 years from now it's going to be 71°F (22°C), you may have to move. (Even if you do not have a problem with the small degree rise in this example, eventually temperatures will rise beyond your tolerance and you can make the calculation for that threshold.) So in 10 years, your 68°F (20°C) summer temperatures will be found a bit farther north, say 20 mi. (32 km), where it is currently around 65°F (18°C) but will similarly warm 3 degrees. That's 20 mi. (32 km) in 10 years, meaning you need to migrate an average of 2 mi. (3.2 km) a year to keep pace with climate change. That distance is not much for a midsized animal, such as a fox or a racoon, especially those who have young every year that must find their own home territories. But what about a slow-growing, sturdy oak? It may need at least 50 years in order to sprout, grow from a flexible sapling to a young, pole-sized tree, and then finally erupt with acorns. And acorns don't fly (though they may hitch a ride with animals). Keeping pace may be considerably more difficult for that crucial component of the ecosystem.

The rate needed to keep pace with climate change has actually been calculated for various scenarios. One that was particularly influential in the scientific community utilized a midrange projection from the Intergovernmental Panel on Climate Change (IPCC), and calculated for every point on the globe how fast you would need to move to keep pace with your current climate (starting in the year 2000). There is a wide range, from around 300 ft. (100 m) a year, the length of a football field or less, to more than 6 mi. (10 km)—quite a trek, especially for small animals or long-lived, slow-growing trees, and probably unrealistic. The areas where the most rapid movement is required are the flat parts of the world: the prairies, steppes, and many deserts. There is no shortcut to a cooler climate in those locations; organisms must move as fast as the climate is warming—and that requisite velocity is likely prohibitive.

In more mountainous terrain, where the option to move up is on the table, distances are much shorter. Once again, landscape heterogeneity is useful. The shortest distances are in the Himalaya, the Rocky Mountains, the Alps, and the Andes. As discussed earlier, we are already seeing that movement as tree line creeps higher around the world, and as plants move into recently deglaciated areas. That migration is happening in real time. But there are limits; mountains have tops. For organisms

LIFE HAS ONLY STARK CHOICES WHEN IT COMES TO A CLIMATE THAT IS LEAVING IT BY THE WAYSIDE: MIGRATE, ADAPT, OR DIE.

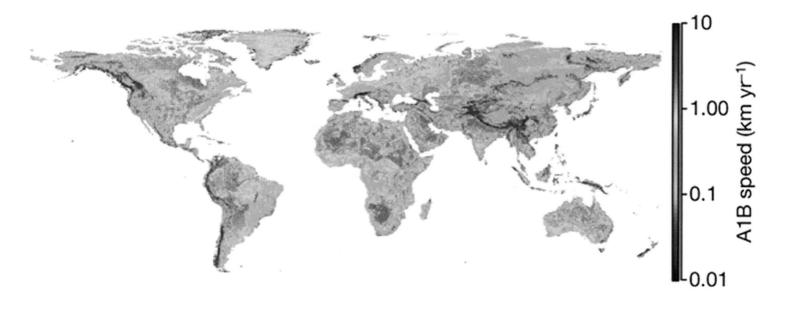

already at the top of a mountain, time is running out quickly. Their options are constrained, because migration off a mountain requires diving down the hills into hotter temperatures. Unless they can muster the fortitude to survive a stretch in an even warmer landscape, they are marooned on a sky island as temperatures rise around them.

Like everything else, the challenges of just keeping pace are not equally distributed around our world. Temperate coniferous forests, like those in the western United States, Europe, and southern Russia, range from only needing to move 109 ft. (100 m) to 6 mi. (10 km) per year, but the vast majority average close to the lesser distance. That comes from their typical distribution on hills and mountainous terrain. Others are not so lucky. The extremely biodiverse mangroves of the world must, on average, move 3280 ft. (1 km) per year, a highly unlikely feat. Essentially, climate will be stable (given its variability) for only about 18 more years for those poor forests. Tropical forests are a bit better off, at about 984 ft. (0.3 km) per year, but that is still a long way considering the years required to germinate, grow, produce seeds, and disperse in the crowded ecosystem that is the tropics.

Once again, Alice's Wonderland is an excellent analogy for our beautiful world beset by chaos and climate confusion. Alice encounters the Red Queen, the ultimate cause of all the trouble for Alice, and they begin to run. But the landscape never

This graphic shows the velocity of climate change, or the estimated distance a species would need to move in a year to maintain a constant climate. An African lizard would need to move up to 10 km poleward every year just to maintain its historical climate (areas in red). Meanwhile, some species in mountainous areas, such as the Himalaya, would still need to move, but much less distance because they can move uphill (at least for a while; those areas are seen in blue). A1B refers to a middle-of-the-road projection in regard to how much warming the world will experience this century.

changes, despite the incredible effort Alice puts into the race. Eventually they stop beneath the same tree under which they started.

> "Well, in our country," said Alice, still panting a little, "you'd generally get to somewhere else—if you run very fast for a long time, as we've been doing."
>
> "A slow sort of country!" said the Queen. "Now, here, you see, it takes all the running you can do, to keep in the same place. If you want to get somewhere else, you must run at least twice as fast as that!"
> —Lewis Carroll, *Through the Looking-Glass*

Indeed, we are not in Alice's country anymore. To keep pace with climate change, life must run very fast just to stay in the same place.

What if it can't? Life could adapt to climate change. But that is quite challenging. At a fundamental level, climate change will generally require evolutionary adaptation (assuming we do not stabilize our emissions soon). This means natural selection for a genetic structure that is more adapted to current conditions (warmer, with more or less precipitation, depending on location) relative to the past genetic structures. Evolution is not conceptually complex; it simply relies on variety in offspring (which always happens—some kids are taller, some shorter, some faster, some slower) and more offspring born than survive to reproduce (the less-adapted individuals are more likely to die). Those that are more "fit" (more likely to have offspring themselves) perpetuate their genes and genetic legacy more successfully than those that are less fit. It is fairly straightforward—have babies with some genetic variation, and that variation influences their ability to survive and pass on their particular combination of genes. While it's not complicated, the mechanisms and patterns are still being unwound. That variation can come from sexual reproduction mixing genes or some recently discovered and still poorly understood mechanisms like epigenetic changes (inheritable changes in phenotype that do not involve DNA changes). The changing climatic conditions, plus that new variability and those new averages, should favor a different set of genes than the previous conditions. But such change takes generations to develop, and generations take time we do not have.

Absent evolution, the other option for species is behavioral change. If it is too hot during the day, forage in the evening (plants don't have the movement option, but they can change things like rooting depths). Desert lizards illustrate this option quite well, because they are already living the reality of the thermal limits on life.

The Husab sand lizard lives in some of the driest and hottest terrain south of the Sahara, the Namib Desert in Africa, where some locations reach 117°F or 47°C. Like many organisms in hot terrain, the lizards cannot tolerate such temperatures for long; their preferred temperature is around 100°F (38°C). At 112°F (44.4°C) the reptiles lose muscle control and can quickly die. So already, the environmental temperature is higher than the lizards can tolerate for some parts of the year, and it has been warming. The lizards have accommodated this behaviorally, finding cooler microclimates hidden within the hillsides and plains, areas under rocks, burrows in the sides of dry streams, or shady spots under bushes. Despite the warming, no populations of the lizard have gone extinct—yet. Their fine-tuned behavioral ways of regulating their internal temperature has allowed them to exploit the few hours a day when temperatures are still amenable to life.

Unfortunately, the climate is still warming. As the average temperature in the Namib inches upward, the thermal window for these creatures to forage is slowly shutting, and eventually will close entirely. Eventually, the time available for the Husab lizards to forage will be too short for them to accumulate enough calories to survive and reproduce, meaning the lizards will either starve in their cooler holes or cook on the sand, the population dwindling. Or a few extremely hot years will bake the sands and kill them off quickly. Unfortunately, behavioral adaptation has its limits as well.

If species cannot migrate or adapt, they will likely die. Species tend to track their climate tolerances, with some variation associated with competition, mutualisms (species interactions that favor both), or some other interactions with other species. In the past, this has meant grand sweeps of life moving toward the equator during cold ice ages and toward the poles during warmer interglacial periods. But as more and more species are outpaced or outmoded by the climates in their current ranges and reaching their limits of adaptation, extinctions will begin in earnest—localized disappearances initially (more affected places first), but eventually true and complete extinctions.

A CAUTIONARY TALE

The first mammal to go globally extinct due to climate change was the Bramble Cay melomys (*Melomys rubicola*), a small rodent endemic to a single island off the coast of Australia, on the northern tip of the Great Barrier Reef. Bramble Cay is some 31 mi. (50 km) off the coast of Papua New Guinea. The island, technically a cay (or a key, as in the Florida Keys), is a small, sandy, very low-lying island (less than 10 ft. or 3 m above sea level) built on the surface of an old coral reef. These islands formed when currents transporting sediment were forced to slow or merge with other currents, dropping their sediment loads to the bottom. The ultimate cause of the melomys's extinction was, in this case, sea level rise. The rodents, which looked like chubby but cheerful rats, lost most of their habitat to the rising high tide line; the overall land surface shrunk by about a third in only 20 years, though it's not entirely gone quite yet. Although we don't know exactly what happened to the last Bramble Cay melomys, it was likely a combination of that lost habitat and a resultant lack of food (meaning lower population size overall), and then frequent high waters that have overrun the island several times in the last few decades. A tropical storm surge, now able to overtop the island because of the higher seas, could have delivered the final knockout. The last one was gone. Thus ended this unique rodent's time on Earth, the first species to become extinct as a direct result of climate change, but likely not the last.

THE ULTIMATE CAUSE OF THE MELOMYS'S EXTINCTION WAS SEA LEVEL RISE.

Although the cause of the melomys extinction is clear (sea level rise reducing habitat and increasing exposure to deadly storm surges), it can be difficult in other cases to directly attribute extinctions to climate change. Most extinctions, either local or global, are attributed to more proximate causes, such as a decrease in food, longer droughts, or altered pathogens—things which themselves are exacerbated by climate change. For example, the largest populations of the Bay checkerspot butterfly (*Euphydryas editha bayensis*), a species endemic to the San Francisco Bay area, have disappeared recently. The survival of these beautiful creatures with striking black, orange, and white wings depends largely on a couple species of host plants (a dwarf plantain, *Plantego erecta*, and to a lesser extent, two species of *Castilleja*) in the area. The larvae of the butterfly must feed on those species and gather enough calories to continue body development. If those plants decline or wither prior to the larvae's feeding period, the butterfly population cannot self-perpetuate. The climate in the area is shifting and the life cycles of these plants no longer synchronize with the butterfly's, a temporal mismatch between the timing of the larvae and the timing of the plants. In addition, the climate is more variable, causing wider population swings. Unfortunately, if a population swings below zero, it never recovers.

There are other proximate factors affecting the Bay checkerspot butterfly. Human development had isolated the populations that were lost, meaning they are unlikely to be recolonized by some surviving population elsewhere. Thankfully, in other locations, topographic variability has so far maintained populations by ensuring some plants and larvae are temporally matched. But the loss of the largest populations presages losses elsewhere. Now, are those losses, both to date and to come, attributable to climate change? The causes are variation and temporal mismatches with a host species, not a direct temperature or precipitation effect—but the relatively recent mismatches are clearly connected to climate. The ultimate cause of many extinctions is the destabilization of relationships between species: their food and hosts, their predators and prey, driven by climate. Even when the final blow is delivered by something else indirectly—a competitor or predator—it does not make extinction any less final.

These localized extinctions are increasing, driven by both climate change and by humans converting land from habitat into farms, homes, and parking lots. The first signs are local extinctions, like the Bay checkerspot butterfly. The loss of a species from an area is not in itself a sign of looming extinction caused by climate change; some populations are inherently unstable and wink in and out of existence, lost due to chance and then recolonized from other populations of the same species that are doing better. And some species simply go extinct for natural reasons. But when those local extinctions start piling up in warmer parts of a species range, or areas where the climate reflects anticipated future conditions, we are right to be worried. A study in 2016 investigated 976 species around the world and found that *47 percent* have seen climate-related local extinctions already, mostly on the warmer and drier edges of their ranges. Local extinctions can presage global extinctions, if they add up. If species cannot migrate to new environments or adapt (or a combination of both), those edge extinctions will continue to nibble away at the overall range until eventually, like the Bramble Cay melomys, the last population is swept away by some chance event.

The unhappy distinction of being the first known mammal to go extinct due to climate change goes to the Bramble Cay melomys. Rising seas chipped away at the animal's habitat, isolated for millennia off the coast of Papua New Guinea. The mylomys population dwindled, individuals finding scant refuge in outcrops of rocks and in the roots of the few plants, until ultimately winking out forever, likely a victim of storms washing over the island. The last known sighting was in 2009.

The Sad Case of the Bramble Cay Melomys

HIGH TIDE AND VEGETATION: 1978–2014

Rising sea levels have slowly overwhelmed the small cay, reducing vegetated area and shrinking the sandy island, resulting in lost habitat and eventual extinction of the melomys.

— High tide line
--- Vegetation margin

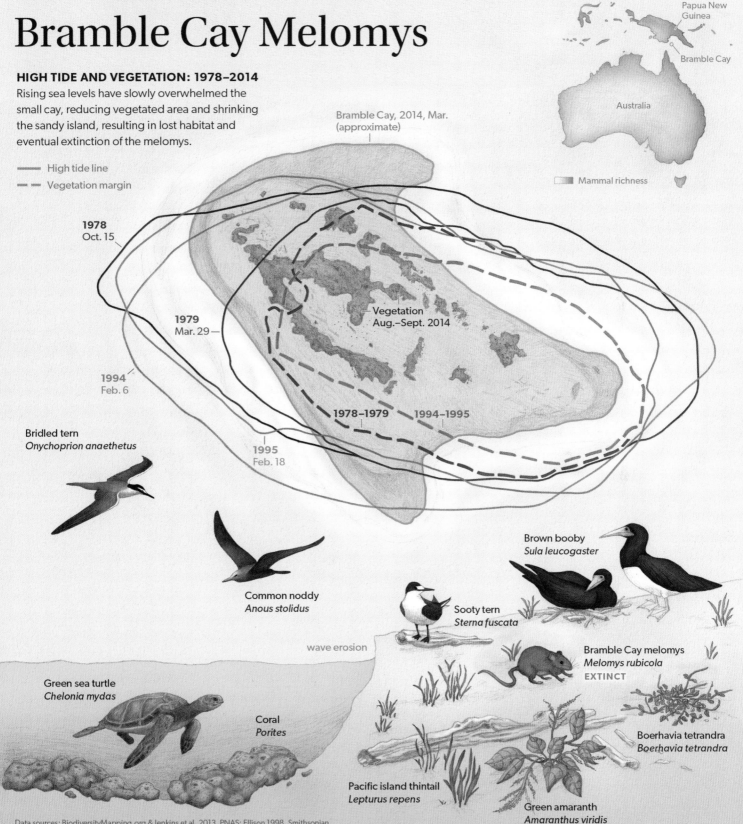

Regional biodiversity hotspots

Papua New Guinea

Bramble Cay

Australia

Mammal richness

Bramble Cay, 2014, Mar. (approximate)

1978 Oct. 15

1979 Mar. 29

1994 Feb. 6

1995 Feb. 18

Vegetation Aug.–Sept. 2014

1978–1979   **1994–1995**

Bridled tern
Onychoprion anaethetus

Common noddy
Anous stolidus

wave erosion

Sooty tern
Sterna fuscata

Brown booby
Sula leucogaster

Green sea turtle
Chelonia mydas

Coral
Porites

Bramble Cay melomys
Melomys rubicola
EXTINCT

Boerhavia tetrandra
Boerhavia tetrandra

Pacific island thintail
Lepturus repens

Green amaranth
Amaranthus viridis

Data sources: BiodiversityMapping.org & Jenkins et al. 2013, PNAS; Ellison 1998. Smithsonian Institution, Washington DC, USA; Gynther et al. 2016. Queensland Government, Brisbane, Australia.

CONCLUSION
THE SCALE OF THE PROBLEM

Exploring Earth's systems like we have in this book can create two, seemingly opposing, reactions. The first is a response of wonder at the variety and range of places, landscapes, and species found on our planet. The variability that defines habitats, and the amazing ways in which life has adapted to that variability, are incredible and arouse a great sense of hope. The heterogeneity of landscapes, from the soil and the canopy to the fractally complex coral reefs, create diversity and life that thrive in their own little pockets. Even the problems associated with urbanity, the pollution and synthetic chemicals, are counterbalanced in some ways by the incredibly interesting habitats humans have created for themselves, with their unique plants and animals, imported foods in the winter, and steady-state indoor climate control. The fact that birds exhibit analogous behavior via migration, following their preferred climates and food sources around the world so they never see a winter, is a fascinating comparison. The interactions, growth patterns, and behaviors that life has engineered to accommodate such variation is complex and intriguing. The scale of those systems dwarfs our human experience. You can discover as much underfoot as you can over the horizon. The human scale of perception is not privileged—we have to look at the world at its proper scale to understand it.

The second response is one of despair. It is impossible to talk about the natural world without talking about anthropogenic climate change, just as it is impossible to talk about biology without evolution. It is the dominant global problem of our age. Climate and its variability (year to year, season to season, day to day) is a foundational reality that all organisms experience, but the current human emission-driven change is not a seasonal phenomenon. It is directional. And it shows no signs of abating. Like a massive battleship that has slowly gotten underway via carbon energy, it is impossible to just row up and push it to a halt. The ship still sways in the waves, sometimes calmer and sometimes faster, but it is moving. And its momentum is monstrous. Understanding and reveling in the first reaction (that the world is a variable place and life has flourished within that variability) means recognizing the second (that changing those variables is bad news for species of all stripes).

The current extinction rate is *hundreds to thousands of times higher* than the historical norm, and the list of species lost forever is long: the dodo bird in 1690 as a result of predation by invasive species; the Steller's sea cow in 1768 as a result of hunting; in 1962, the Hawaiian chaff flower because of habitat loss from military bases; in 2003, the St. Helena olive because of timber harvesting and habitat loss. The effects of climate change are now adding to those other, more localized, human factors, and climate is a much more pervasive foe. The atmosphere is a common pool, and we are ruining it.

The charts and maps that accompany this narrative, forming a graphic way of viewing the natural world through imagery, brutally illustrate the challenge facing our world due to climate change. They illustrate the scape and scope of a problem disconnected from our everyday experience, but all the more dangerous for that fact.

Each chart is also a snapshot of conditions, a point of time or a frame of reference. The historical imagery captures the world as it was, or as it was perceived to

YOU CAN DISCOVER AS MUCH UNDERFOOT AS YOU CAN OVER THE HORIZON.

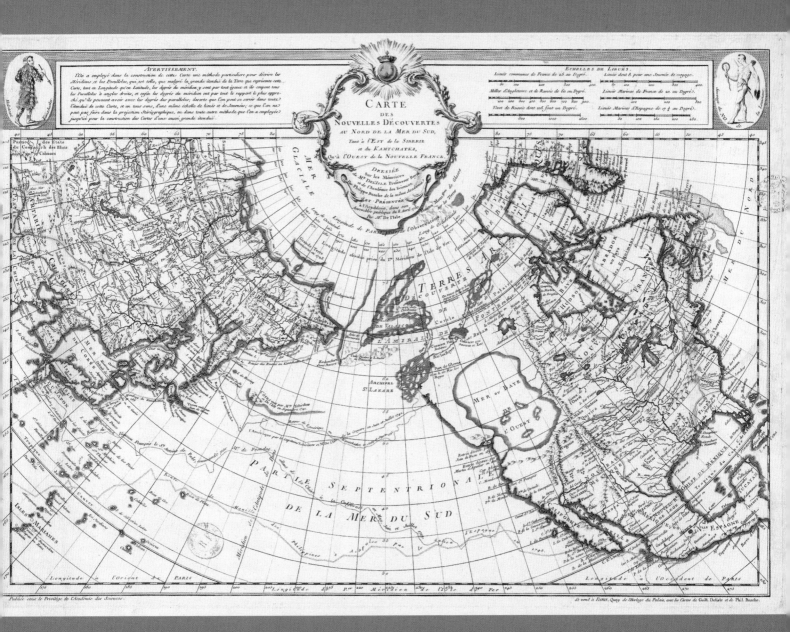

This 1752 map of the Pacific Ocean compiled by Philippe Buache, perhaps the most extensively notated of its kind, visually conveys the excitement of exploration that permeated the early scientific cartographers (the text is in French). The Pacific Coast of North America is distorted nearly beyond recognition, a result of infrequent mapping and poor locational accuracy. As our skill in mapping has improved, so has our understanding of the complex mechanisms across space and time that shape our natural world.

MAPP
Dreßé fu
de M.rs de l'Acad
et quelques autres e

SON ALT
MON
LE DUC

Pole Arctique

Cercle Polaire

TERRES ARCTIQUES
GROENLAND

ISLANDE
Islande

Frislande
Cap de Farvel

Nec Ultra
Port de Munck

BAYE D'HUDSON

Echurchu
Fort Bourbon

CANADA ou NOUVELLE FRANCE

C. Mendocino
C. de Fortune

CALIFORNIE

NOUVEAU MEXIQUE

C. de S. Lucas

AMERIQUE SEPTENTRIONALE
FLORIDE

GOLFE DE MEXIQUE

MER DU SUD ou MER PACIFIQUE

MER DU NORD

Tropique du Cancer

ISLES ACORES

MER DE SARGASSE

ISLES DU CAP VERD

NOUVELLE ESPAGNE

Equateur ou La Ligne

Isles Gualapegas

I. DE SALOMON

Terre que la flote de Mendaña crut être la N.le Guinée

la Solitaire

TERRE FERME
GUIANE

PAYS DES AMAZONES

AMERIQ. MERIDIONALE

BRESIL

Tropique du Capricorne

N.lle ZELANDE

TERRE MAGELLANIQUE

TERRES AUSTRALES

Cercle Polaire

Pole Antarctique

A
Chez
des Car
Sulpi
Avec
pou

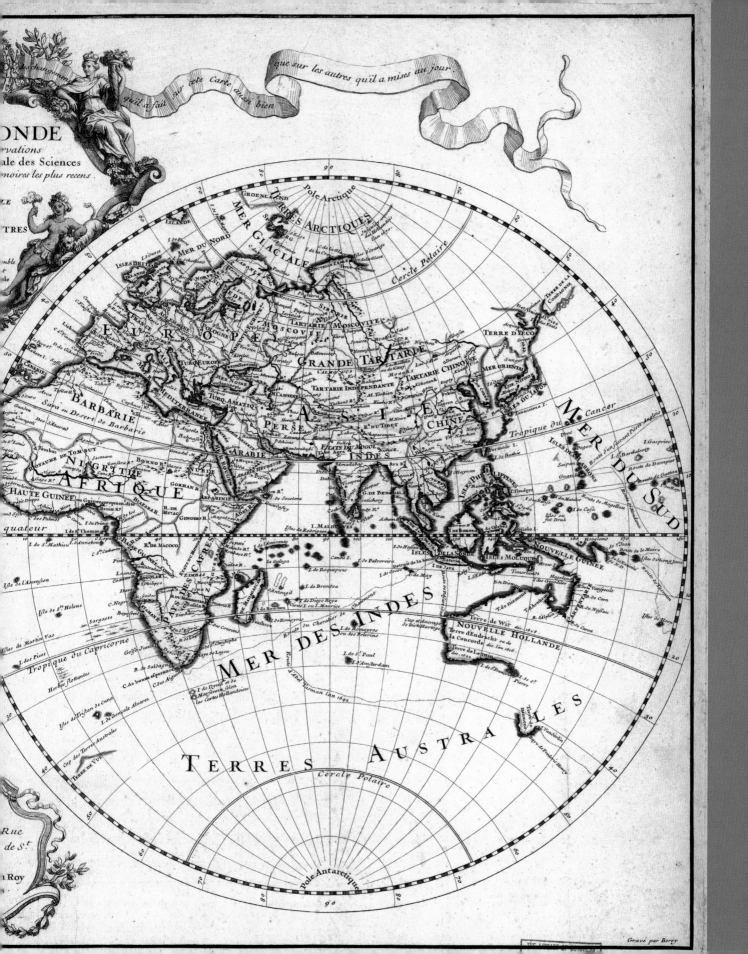

ONDE
rvations
ale des Sciences
moires les plus recens

qu'il a fait sur cette Carte que sur les autres qu'il a mises au jour

Pole Arctique

TERRES ARCTIQUES

GROENLANDE

MER GLACIALE

ISLANDE

Cercle Polaire

MER DU NORD

EUROPE

POLOGNE

MOSCOVIE

SIBERIE

TARTARIE MOSCOVITE

TERRE D'YECO

GRANDE TARTARIE

TARTARIE CHINOISE

MER ORIENTALE

BARBARIE

MEDITERRANEE

TURQ: ASIATIQ

PERSE

ASIE

CHINE

MER DU JAPON

MER DU SUD

Sara ou Desert de Barbarie

ARABIE

ÉTATS DU MOGOL

INDES

Cancer

Tropique du

ISLES DES LARRONS

ROYAUME DE TOMBUT

NIGRITIE

AFRIQUE

ARABIE HEUREUSE

I. MALDIVES

G. DE BENGALE

ISLE PHILIPPINES

I. Bartholemy

HAUTE GUINEE

ABISSINIE

Formose I.

Equateur

I. de St. Thomas

I. DE BORNEO

Gilolo I.

I. de St. Mathieu

Annobon

ISLES DE LA SONDE

ISLES MOLUQUES

NOUVELLE GUINEE

BASSE GUINEE

ZIMBAS

I. DE JAVA

CONGO

ANGOLA

PAYS DES CAFRES

MONOMOTAPA

MADAGASCAR

Terre de Wit

NOUVELLE HOLLANDE

Terre d'Endracht ou de la Concorde

MER DES INDES

Tropique du Capricorne

B. de Saldagne

C. de bonne esperance

C. des Aiguilles

I. d'Amsterdam

I. de St. Paul

TERRES   AUSTRALES

Cercle Polaire

Pole Antarctique

Gravé par Berey

PREVIOUS An early and beautiful map created by Guillaume de L'Isle (cartographer) and Claude Berey (engraver) in 1700 shows early exploration routes and contains extensive secondhand notes from a variety of sources. This map illustrates how the world was continually expanding for early explorers, and with it our understanding of the scope of natural processes. Among many other innovations, de L'Isle's map appears to have been the first to correctly place California on the North American continent (rather than depicting it as an island). He is sometimes considered the first scientific cartographer, thanks to his attention to details like correct latitudes and spatial relationships.

CONCLUSION

246

be: a source of food, an obstacle to navigate around, or a challenge to be understood. The earliest scientific maps opened our eyes to the scale of the world, the extent of a hurricane or the diversity of land within a mountain range. Later visualizations attempted to display more complex phenomena: the diversity of species interacting in a city, the ways in which ice flows downhill, or the shape of a migration route that spans a continent.

If we are not careful, however, these snapshots can also lure us into thinking the world is made of averages and is static. Habitat maps can be falsely encouraging by showing that even for threatened species, some land remains, not all is lost! The Bramble Cay melomys had a little habitat left, too. But it wasn't enough to hide from climate change. The future is a moving target, and there is little evidence to suggest that the world will stop changing. So, are the charts only documenting a decline, a loss of sensitive species, and the general homogenization of a world in which only heat-tolerant species adapt and mountaintops are populated by climate refugees? In a changing climate, maps and images are outdated the moment they are printed. Models and maps for the future are just as problematic, as the world evolves and renders the underlying datasets for the projections obsolete.

NO PIECE IS SEPARATE FROM THE WHOLE.

The best defense is an understanding of basic principles, the same knowledge that inspired Humboldt to describe the world in a new way and that led Arrhenius to connect climate change with CO_2 emissions in 1896. Understanding how the atmosphere works leads immediately to the obvious threat posed by carbon emissions. Understanding what creates grasslands instead of forests leads immediately to recognizing the threats posed by more precipitation or less fire. A small grasp of chemistry leads to an immediate understanding of the potential problems associated with ocean acidification. The ecological functioning of cities reveals a lot about why we have them in the first place. The details are complicated, keeping scientists busy pulling all the threads and following all the complicated interactions that result in the final result. The proximate drivers are complicated, but the ultimate drivers—physics, chemistry, climate—are still edifying.

"Mother Nature is only chemistry, biology and physics and responds to nothing else," wrote Thomas Friedman in the *New York Times* on June 16, 2020, in an opinion piece about COVID-19, the novel coronavirus that began circulating in 2019. Nearly this entire book was written in quarantine. While the virus itself is not a human creation, it flourishes in the habitat we have created; it is a creature that thrives on density. Something at the microscopic scale, too small to see, has ground the human world to a halt, and thereby influenced global environmental systems. Small things can enact change at a big scale. Global air pollution declined as a result of decreased economic activity, but human suffering increased in equal measure. We have remade this planet in our own energy-hungry, density-loving image, and we must own that fact and how it has influenced the rest of our compatriots—plants, animals, mountains, seas, and others—on this small, blue marble. The existential challenges facing the world as I write this are immense. No piece is separate from the whole and the whole is in trouble. But one has to believe there is hope. After all, we are part of that biology ourselves.

SUGGESTED READING

Included here are select readings that can support the details covered in this book, for interested readers who want to dive deeper into a particular topic. The list is not exhaustive but should provide some interesting supplements and thought-provoking material. Several are websites, which may unfortunately go out of date, but can offer unique data.

Many of the following are references to primary scientific literature, meaning peer-reviewed publications in which our scientific knowledge is first disseminated. That can make them seem opaque to the nonspecialist, as they are written primarily for brevity and data communication, not readability. Sometimes they may even appear to contradict each other—but that is not a bad thing, merely a natural consequence of building knowledge. Data is always incomplete at the beginning (just as it was 100 years ago). Arguments about interpretations or significance ensue. That is the process of science.

I encourage you to check out the primary literature yourself. Much may be confusing. Do not hesitate to skim, take what you can, and then to contact the authors of the papers. They will be happy to hear from you and explain any sticking points. (Contact information for the corresponding author will always be on the first page of the article.)

The immense world of scientific knowledge is out there, at your fingertips, thanks to the internet. No subscription is required (and even if there is, you can contact the author to ask questions directly).

Enjoy!

INTRODUCTION

The problem of scale is not just in understanding, but also in management, even when the process is understood. For a semi-technical discussion:

Cumming, G. S., D. H. Cumming, and C. L. Redman. "Scale Mismatches in Social-Ecological Systems: Causes, Consequences, and Solutions." *Ecology and Society* 11, no. 1 (2006): 14.

For a discussion of why our human brain is poorly adapted to conceptualizing climate change—or really, any long-term environmental system that operates on the scale of decades to centuries (or longer):

Pahl, S., S. Sheppard, C. Boomsma, and C. Groves. "Perceptions of Time in Relation to Climate Change." *Wiley Interdisciplinary Reviews: Climate Change* 5, no. 3 (2014): 375–388.

Mapping helps the scale mismatch, as discussed via the example of Alexander von Humboldt. He was quite a celebrity, and his impact came as much from his ability to capture the popular imagination as it did from his scientific findings:

Schaumann, C. "Who Measures the World? Alexander von Humboldt's Chimborazo Climb in the Literary Imagination." *The German Quarterly* 82, no. 4 (2009): 447–468.

For the full scientific conversation about the Chimborazo resurvey mentioned in the text:

Morueta-Holme, N., K. Engemann, P. Sandoval-Acuña, J. D. Jonas, R. M. Segnitz, and J. C. Svenning. "Strong Upslope Shifts in Chimborazo's Vegetation over Two Centuries since Humboldt." *Proceedings of the National Academy of Sciences* 112, no. 41 (2015): 12741–12745.

Sklenář, P. "Advance of Plant Species on Slopes of the Chimborazo Volcano (Ecuador) Calculated Based on Unreliable Data." *Proceedings of the National Academy of Sciences* 113, no. 4 (2016): E407–E408.

Morueta-Holme, N., K. Engemann, P. Sandoval-Acuña, J. D. Jonas, R. M. Segnitz, and J. C. Svenning. "Resurvey of Antisana Supports Overall Conclusions of Chimborazo Study." *Proceedings of the National Academy of Sciences* 116, no. 43 (2019): 21346–21347.

Moret, P., P. Muriel, R. Jaramillo, and O. Dangles. "Reply to Morueta-Holme et al.: Humboldt's Historical Data Are Not Messy, They Just Need Expert Examination." *Proceedings of the National Academy of Sciences* 116, no. 43 (2019): 21348–21349.

Morueta-Holme, N., K. Engemann, P. Sandoval-Acuña, J. D. Jonas, R. M. Segnitz, and J. C. Svenning. "Reply to Sklenář: Upward Vegetation Shifts on Chimborazo Are Robust." *Proceedings of the National Academy of Sciences* 113, no. 4 (2016): E409–E410.

The Intergovernmental Panel on Climate Change has the most rigorous discussion of environmental conditions associated with climate change, including carefully vetted statements about scale and uncertainty. Nonscientists should focus on the "Summary for Policy Makers," which is explicitly designed to communicate the state of the science for a variety of Earth systems:

https://www.ipcc.ch/report/sr15/summary-for-policymakers/.

ATMOSPHERE

A classic text highlighting how early scientists struggled with mapping phenomena much bigger than their instruments could handle, at least individually:

Hayden, Edward Everett. *The Modern Law of Storms.* New York City: L. R. Hamersly & Co., 1890.

For more reading on the famous Great White Hurricane of 1888, explore the New England Historical Society:

https://www.newenglandhistoricalsociety.com/great-white-hurricane-of-1888/.

And a re-analysis:

Kocin, P. J. "An Analysis of the "Blizzard of '88." *Bulletin of the American Meteorological Society* 64, no. 11 (1983): 1258–1272.

The fascinating story of an important metric of global atmospheric carbon dioxide called the Keeling Curve is nicely summarized by a University of California article:

Monroe, R. "The History of the Keeling Curve." *The Keeling Curve Blog*, April 3, 2013. University of California San Diego and Scripps Institution of Oceanography. https://sioweb.ucsd.edu/programs/keelingcurve/2013/04/03/the-history-of-the-keeling-curve/.

The groundbreaking work of Swedish Nobel Prize winner Svante Arrhenius that first tied greenhouse gases to a changing climate can be found here:

Arrhenius, S. "On the influence of carbonic acid in the air upon the temperature of the ground." *The London, Edinburgh, and Dublin Philosophical Magazine and Journal of Science* 41, no. 251 (1896): 237–276.

Glacier Bay is also an incredible story. For the human history:

Connor, C., G. Streveler, A. Post, D. Monteith, and W. Howell. "The Neoglacial Landscape and Human History of Glacier Bay, Glacier Bay National Park and Preserve, Southeast Alaska, USA." *The Holocene* 19, no. 3 (2009): 381–393.

For the story of vegetation change over the last century, the most up-to-date summary is here:

Buma B., S. M. Bisbing, G. Wiles, and A. L. Bidlack. "100 Years of Primary Succession Highlights Stochasticity and Competition Driving Community Establishment and Stability." *Ecology* 100, no. 12 (2019).

A more accessible version of the above successional story:

Buma B., S. Bisbing, J. Krapek, and G. Wright. "A Foundation of Ecology Re-Discovered: 100 Years of Succession on the William S Cooper Permanent Plots Shows Importance of Contingency in Community Development." *Ecology* 98, no. 6 (2017): 1513–1523.

A period called the Little Ice Age does not yet have a truly comprehensive explanation for its onset, duration, and extent. This older book does a good job with the basic evidence and patterns found around the world:

Grove, Jean M. 1988. *The Little Ice Age*. Metheun and Company Ltd., Taylor & Francis Group. Republished in the Taylor and Francis e-Library 2008.

A paper on the volcanic hypothesis for the Little Ice Age, one of the possible causes currently seeing strong evidence:

Miller, G. H., Á. Geirsdóttir, Y. Zhong, D. J. Larsen, B. L. Otto Bliesner, M. M. Holland, D. A. Bailey, K. A. Refsnider, S. J. Lehman, J. R. Southon, and C. Anderson. "Abrupt Onset of the Little Ice Age Triggered by Volcanism and Sustained by Sea Ice/Ocean Feedbacks." *Geophysical Research Letters* 39, no. 2 (2012).

While human impacts were likely only a minor component of the Little Ice Age, it is a fascinating story:

Koch, A., C. Brierley, M. M. Maslin, and S. L. Lewis. "Earth System Impacts of the European Arrival and Great Dying in the Americas after 1492." *Quaternary Science Reviews* 207 (2019): 13–36.

OCEAN

An investigation into sea level rise and impacts on coastal cities:

Strauss, B. H., S. Kulp, and A. Levermann. "Mapping Choices: Carbon, Climate, and Rising Seas, Our Global Legacy." Report in *Climate Central* (2015): 1–38.

And the same, for coastal protected areas in Africa—an under-studied continent with regard to sea level rise:

Brito, J. C., and M. Naia. "Coping with Sea-Level Rise in African Protected Areas: Priorities for Action and Adaptation Measures." *BioScience* 70, no. 10 (2020): 924–932.

How modern elevation science is refining our expectations for sea level rise:

Kulp, S. A., and B. H. Strauss. "New Elevation Data Triple Estimates of Global Vulnerability to Sea-Level Rise and Coastal Flooding." *Nature Communications* 10, no. 1 (2019): 1–12.

The Intergovernmental Panel on Climate Change has an excellent discussion of the complexities of sea level rise; see Table 4.1 for explicit statistics:

https://www.ipcc.ch/srocc/chapter/chapter-4-sea-level-rise-and-implications-for-low-lying-islands-coasts-and-communities/.

Figuring out the role of heat in the ocean is a challenge, and scientists are continually refining estimates and improving models of what really happens in the oceans:

Zanna, L., S. Khatiwala, J. M. Gregory, J. Ison, and P. Heimbach. "Global Reconstruction of Historical Ocean Heat Storage and Transport."

Proceedings of the National Academy of Sciences 11, No. 64 (2019): 1126–1131.

Florida has a fascinating geological history, which leads to its sensitivity to sea level rise:
Milton, C., and R. Grasty. "'Basement' Rocks of Florida and Georgia." *AAPG Bulletin* 53, No. 12 (1969): 2483–2493.

For a discussion on the social ramifications, Yale University published an illuminating piece online:
https://e360.yale.edu/features/as-miami-keeps-building-rising-seas-deepen-its-social-divide.

Local projections in Florida (compare to previously noted IPCC reports to see the challenges of talking about phenomena at one scale versus another, and the challenges of making predictions when science itself is still learning about the mechanisms of ocean heating, expansion, and overall rise):
"Unified Sea Level Rise Projection Southeast Florida." Document prepared by the Southeast Florida Regional Climate Change Compact's Sea Level Rise Ad Hoc Work Group. (2019) https://southeastfloridaclimatecompact.org/wp-content/uploads/2020/04/Sea-Level-Rise-Projection-Guidance-Report_FINAL_02212020.pdf.

Shanghai has similar challenges, and several engineering plans to address them:
Lu M., and J. Lewis. "Shanghai: Targeting Flood Management," *in China and US Case Studies: Preparing for Climate Change,* a six-report series. Washington DC: Georgetown Climate Center, 2015. https://www.georgetownclimate.org/files/report/GCC-Shanghai_Flooding-August2015.pdf.

Ocean acidification is a fascinating and deep topic. A brief primer from the US National Oceanic and Atmospheric Administration and one from Woods Hole Oceanographic Institute are good starting points for the unfamiliar:
https://pmel.noaa.gov/CO₂/story/A+primer+on+pH.
https://www.whoi.edu/oceanus/feature/ocean-acidification-a-risky-shell-game/.

Two more detailed discussions of acidification:
Doney, S. C., V. J. Fabry, R. A. Feely, and J. A. Kleypas. "Ocean Acidification: The Other CO₂ Problem." *Annual Review of Marine Science* 1, no. 1 (2009): 169–192.
Strong, A. L., K. J. Kroeker, L. T. Teneva, L. A. Mease, and R. P. Kelly. "Ocean Acidification 2.0: Managing our Changing Coastal Ocean Chemistry." *BioScience* 64, no. 7 (2014): 581–592.

Finding the world's longest river:
Liu, S., P. Lu, D. Liu, P. Jin, and W. Wang. "Pinpointing the Sources and Measuring the Lengths of the Principal Rivers of the World." *International Journal of Digital Earth* 2, no. 1 (2009): 80–87.

FRESH WATER

For centuries, rivers have been a draw for civilizations. This paper is one of the first to explore the strength of that draw and how it is disappearing—a fundamental shift in humanity's relationship with the natural world, and one that is worrisome for water experts:

Fang, Y., and J. W. Jawitz. "The Evolution of Human Population Distance to Water in the USA from 1790 to 2010." *Nature Communications* 10, no. 1 (2019): 1–8.

The science of atmospheric rivers:

Ralph, F. M., and M. D. Dettinger. "Storms, Floods, and the Science of Atmospheric Rivers." *Eos, Transactions, American Geophysical Union* 92, no. 32 (2011): 265–266.

The below exploration of the meteorology of atmospheric rivers is dense, but attempts to address an all-important question: Are they getting more frequent? Or less?

Dettinger, M. D., F. M. Ralph, and J. J. Rutz. "Empirical Return Periods of the Most Intense Vapor Transports during Historical Atmospheric River Landfalls on the US West Coast." *Journal of Hydrometeorology* 19, no. 8 (2018): 1363–1377.

The story of the 1927 floods is told in text and music. To add to the below documents, look for recordings from Memphis Minnie and others about the monumental event:

Smith, J. A., and M. L. Baeck. "'Prophetic Vision, Vivid Imagination': The 1927 Mississippi River Flood." *Water Resources Research* 51, no. 12 (2015): 9964–9994.

Henry, A. J. "Frankenfield on the 1927 Floods in the Mississippi Valley." *Monthly Weather Review* 551 (1927): 437-452. https://doi.org/10.1175/1520-0493 (1927)55<437:FOTFIT>2.0.CO;2.

The biggest lake west of the Mississippi, now lost to legend and farmland:

Preston, W. L. "The Tulare Lake Basin: An Aboriginal Cornucopia." *The California Geographer* 30 (1990).

Haslam, G. "The Lake That Will Not Die." *California History* 72, no. 3 (1993): 256–271.

Land subsidence in California:

https://ca.water.usgs.gov/land_subsidence/california-subsidence-areas.html.

The story of salmon resilience in the California water network:

Phillis, C. C., A. M. Sturrock, R. C. Johnson, and P. K. Weber. "Endangered Winter-Run Chinook Salmon Rely on Diverse Rearing Habitats in a Highly Altered Landscape." *Biological Conservation* 217 (2018): 358–362.

WETLANDS

Wetland trends around the world are followed through what is, essentially, an accounting project, and worth tracking for these essential land cover types.

Davidson, N. C. "How Much Wetland Has the World Lost? Long-Term and Recent Trends in Global Wetland Area." *Marine and Freshwater Research* 65, no. 10 (2014): 934–941.

Reis, V., V. Hermoso, S. K. Hamilton, D. Ward, E. Fluet-Chouinard, B. Lehner, and S. Linke. "A Global Assessment of Inland Wetland Conservation Status." *BioScience* 67, no. 6 (2017): 523–533.

The story of malaria is an unexpected component of our longstanding, divided relationship with wetlands:

Rosenthal, P. J., C. C. John, and N. R. Rabinovich. "Malaria: How Are We Doing and How Can We Do Better?" *The American Journal of Tropical Medicine and Hygiene* 100, no. 2 (2019): 239.

Rey, J. R., W. E. Walton, R. J. Wolfe, C. R. Connelly, S. M. O'Connell, J. Berg, G. E. Sakolsky-Hoopes, and A. D. Laderman. "North American Wetlands and Mosquito Control." *International Journal of Environmental Research and Public Health* 9, no. 12 (2012): 4537–4605.

Wetlands in Alaska:

Clewley, D., J. Whitcomb, M. Moghaddam, K. McDonald, B. Chapman, P. Bunting, "Evaluation of ALOS PALSAR Data for High-Resolution Mapping of Vegetated Wetlands in Alaska." *Remote Sensing* 7 (2015): 7272–7297.

FORESTS

Humans have long managed forests intensively— longer than most people realize. The following three sources discuss forest management in the Amazon and northwestern North America as a start:

Piperno, D. R., C. N. McMichael, and M. B. Bush. "Finding Forest Management in Prehistoric Amazonia." *Anthropocene* 26 (2019): 100211.

Armstrong, C. G. D. "Historical Ecology of Cultural Landscapes in the Pacific Northwest." PhD dissertation, Simon Fraser University, 2017.

Storm, L., and D. Shebitz. "Evaluating the Purpose, Extent, and Ecological Restoration Applications of Indigenous Burning Practices in Southwestern Washington." *Ecological Restoration* 24, no. 4 (2006): 256–268.

While we can map forest loss well with satellites, saying why the forest was lost is another matter:

Curtis, P. G., C. M. Slay, N. L. Harris, A. Tyukavina, and M. C. Hansen. "Classifying Drivers of Global Forest Loss." *Science* 361, no. 6407 (2018): 1108–1111.

The shifting climate means increasing stresses on forests, many of which will look resilient and tough—until they are not:

Daley, J. "California's Drought Killed Almost 150 Million Trees." smithsonianmag.com, July 10, 2019. https://www.smithsonianmag.com/smart-news/why-californias-drought-killed-almost-150-million-trees-180972591/.

MOUNTAINS

How altitude is similar to latitude:

Montgomery, K. "Variation in Temperature with Altitude and Latitude." *Journal of Geography* 105, no. 3 (2006): 133–135.

A discussion of what may be one of the oldest spatial representations created by people, on mammoth tusks:

Svoboda, J. "On Landscapes, Maps and Upper Paleolithic Lifestyles in the Central European Corridor: the Images of Pavlov and Predmostí." *Veleia* 34, no. 1 (2017): 67–74.

Mountain glaciers contribute directly to sea level rise, because they are completely removed from the oceanic system (unlike oceanic ice). Quantifying their contribution is an ongoing challenge:

Zemp, M., M. Huss., E. Thibert, N. Eckert, R. McNabb, J. Huber, M. Barandun, H. Machguth, S. U. Nussbaumer, I. Gärtner-Roer, and L. Thomson. "Global Glacier Mass Changes and Their Contributions to Sea-Level Rise from 1961 to 2016." *Nature* 568, no. 7752 (2019): 382–386.

IPCC Special Report on High Mountain Areas: https://www.ipcc.ch/srocc/chapter/chapter-2/.

Humans live in these landscapes, and adaptation is a challenge, especially when the environment is so extreme to begin with:

Gentle, P., and T. N. Maraseni. "Climate Change, Poverty and Livelihoods: Adaptation Practices by Rural Mountain Communities in Nepal." *Environmental Science & Policy* 21 (2012): 24–34.

For more on snowpack trends in the Western US:

Mote, P. W., S. Li., D. P. Lettenmaier, M. Xiao, and R. Engel. "Dramatic Declines in Snowpack in the Western US." *npj Climate and Atmospheric Science* 1 (2018): 1–6.

The original description of the US Mountain West by the explorer and gifted writer, John Wesley Powell, this book talks of his three-month exploration of the Green and Colorado Rivers, including the first push through the Grand Canyon by an American. Several editions are available of this classic work.

Powell, J. W. *The Exploration of the Colorado River and its Canyons*. New York City: Penguin Classics, 2003.

GRASSLANDS

Modern science and management is still grappling with the immense but fragile resource of limited water in a dry but potentially productive land:

Hornbeck, R., and P. Keskin. "The Historically Evolving Impact of the Ogallala Aquifer: Agricultural Adaptation to Groundwater and Drought." *American Economic Journal: Applied Economics* 6, no. 1 (2014): 190–219.

The original optimism associated with water in the Ogallala Aquifer:

Weakly, H. E., and L. L. Zook. "Pump Irrigation Results (North Platte Substation)." *University of Nebraska Experiment Station Bulletin* 227 (June, 1928).

The story of the 100th meridian builds from the constraints of global and regional climate to make a distinct, and still present, human fingerprint on the land:

Seager, R., J. Feldman, N. Lis, M. Ting, A. P. Williams, J. Nakamura, H. Liu, and N. Henderson. "Whither the 100th Meridian? The Once and Future Physical and Human Geography of America's Arid–Humid Divide. Part II: The Meridian Moves East." *Earth Interactions* 22, no. 5 (2018): 1-24.

On the evolution of the Ogallala and Great Plains aquifer system over geological time:

Willett, S. D., S. W. McCoy, and H. W. Beeson. "Transience of the North American High Plains Landscape and its Impact on Surface Water." *Nature* 561, no. 7724 (2018): 528–532.

TUNDRA

More information on the feedback between permafrost thaw, carbon release, and climate change:

Turetsky, M. R., B. W. Abbott, M. C. Jones, K. Walter Anthony, D. Olefeldt, E. A. G. Schuur, C. Koven, A. D. McGuire, G. Grosse, P. Kuhry, G. Hugelius, D. M. Lawrence, C. Gibson, A. B. K. Sannel. "Permafrost Collapse is Accelerating Carbon Release." *Nature* 569, no. 7754 (2019): 32–34.

Schuur, E. A. G., A. D. McGuire, C. Schädel, G. Grosse, J. W. Harden, D. J. Hayes, G. Hugelius, C. D. Koven, P. Kuhry, D. M. Lawrence, S. M. Natali, D. Olefeldt, V. E. Romanovsky, K. Schaefer, M. R. Turetsky, C. C. Treat, and J. E. Vonk. "Climate Change and the Permafrost Carbon Feedback." *Nature* 520 (2015): 171–179.

For more information on the role of climate warming on tundra plants:

Elmendorf, S. C., G. H. Henry, R. D. Hollister, R. G. Björk, A. D. Bjorkman, T. V. Callaghan, L. S. Collier, E. J. Cooper, J. H. Cornelissen, T. A. Day, and A. M. Fosaa. "Global Assessment of Experimental Climate Warming on Tundra Vegetation: Heterogeneity Over Space and Time." *Ecology Letters* 15, no. 2 (2012): 164–175.

Mapping thermokarst features:

Olefeldt, D., S. Goswami, G. Grosse, D. Hayes, G. Hugelius, P. Kuhry, A. D McGuire, V. E. Romanovsky, A. B. K. Sannel, E. A. G. Schuur, and M. R. Turetsky. "Circumpolar Distribution and Carbon Storage of Thermokarst Landscapes." *Nature Communications* 7, no. 1 (2016): 1–11.

The role of snow, and loss of snow, in the tundra:

Sturm, M., J. Holmgren, J. P. McFadden, G. E. Liston, F. S. Chapin III, and C. H. Racine. "Snow–Shrub Interactions in Arctic Tundra: A Hypothesis with Climatic Implications." *Journal of Climate* 14, no. 3 (2001): 336–344.

DESERTS

A primer on desert life around the world:

Ward, D. *The Biology of Deserts*. Oxford University Press, 2nd edition, 2016.

Documentation of the fascinating desert adaptations found in one of the most arid deserts in the world, Saudi Arabia:

Petrie, J. M. "Arabian Desert Primer: Ornamental Potential of Hyper-Arid Adapted Plants from Saudi Arabia." *Desert Plants* 23, no. 1 (2007): 19–32.

Adaptations in desert plants go beyond saving water, to questions of energetics:

Gibson, A. C. "Photosynthetic Organs of Desert Plants." *BioScience* 48, no. 11 (1998): 911–920.

URBAN

A review of human-created soils, and official classification scheme:

Galbraith, J., and R. K. Shaw. "Human-Altered and Human Transported Soils." In USDA *Soil Survey Manual*, USDA *Handbook 18*, edited by C. Ditzler, K. Scheffe, and H.C. Monger. Washington DC: US Government Printing Office, 2017. https://www.nrcs.usda.gov/wps/portal/nrcs/detail/soils/ref/?cid=nrcseprd1343023.

The story of plaggen soils, one of the oldest human-created soil types:

Blume, H. P., and P. Leinweber. "Plaggen Soils: Landscape History, Properties, and Classification." *Journal of Plant Nutrition and Soil Science* 167, no. 3 (2004): 319–327.

On light pollution in urban areas:

Falchi, F., P. Cinzano, D. Duriscoe, C. C. Kyba, C. D. Elvidge, K. Baugh, B. A. Portnov, N. A. Rybnikova, and R. Furgoni. "The New World Atlas of Artificial Night Sky Brightness." *Science Advances* 2, no. 6 (2016): e1600377.

Urban areas require energy inputs, which is calculated as the appropriation of net primary productivity from plants:

Krausmann, F., K. H. Erb, S. Gingrich, H. Haberl, A. Bondeau, V. Gaube, C. Lauk, C. Plutzar, and T. D. Searchinger. "Global Human Appropriation of Net Primary Production Doubled in the 20th Century." *Proceedings of the National Academy of Sciences* 110, no. 25 (2013): 10324–10329.

Litchfield, Connecticut, history, with oral history inclusions:

White, A. C. *The History of the Town of Litchfield, Connecticut, 1720-1920.* Litchfield, Connecticut: Enquirer print., 1920.

On the cherry trees of Washington DC:

US National Park Service. "Eliza Scidmore's Faithful Pursuit of a Dream." https://www.nps.gov/articles/scidmore.htm.

A full categorization of Great Barrier Reef fauna:
https://www.gbrmpa.gov.au/the-reef/animals.

Historical work from Adelaide, Australia, documenting the changes in animals over the process of urbanization:

Tait, C. J., C. B. Daniels, and R. S. Hill. "Changes in Species Assemblages Within the Adelaide Metropolitan Area, Australia, 1836–2002." *Ecological Applications* 15, no. 1 (2005): 346–359.

Urban biodiversity compared to rural areas in Europe:

Ferenc, M., O. Sedláček, R. Fuchs, M. Dinetti, M. Fraissinet, and D. Storch. "Are Cities Different? Patterns of Species Richness and Beta Diversity of Urban Bird Communities and Regional Species Assemblages in Europe." *Global Ecology and Biogeography* 23, no. 4 (2014): 479–489.

On urban areas and their effect on the biodiversity of birds and trees:

Chace, J. F., and J. J. Walsh. "Urban Effects on Native Avifauna: A Review." *Landscape and Urban Planning* 74, no. 1 (2006): 46–69.

Alvey, A. A. "Promoting and Preserving Biodiversity in the Urban Forest." *Urban Forestry & Urban Greening* 5, no. 4 (2006): 195–201.

LIFE

How fast species need to move to keep pace with their (historical) average climatic conditions:

Loarie, S. R., P. B. Duffy, H. Hamilton, G. P. Asner, C. B. Field, and D. D. Ackerly. "The Velocity of Climate Change." *Nature* 462, no. 7276 (2009): 1052–1055.

How, and how much, protected areas can help:

Alway, J., L. Moyer-Horner, and M. B. Palamar. "Climate Change and Species Range Dynamics in Protected Areas." *BioScience* 61, no. 10 (2011): 752–761.

More on the sad case of the Bramble Cay melomys:

Gynther, I., N. Waller, and L. K.-P. Leung. "Confirmation of the Extinction of the Bramble Cay Melomys *Melomys rubicola* on Bramble Cay, Torres Strait: Results and Conclusions from a Comprehensive Survey in August–September 2014." Unpublished report to the Department of Environment and Heritage Protection, Queensland Government, Brisbane, Australia. 2016.

Waller, N. L., I. C. Gynther, A. B. Freeman, T. H. Lavery, and L. K.-P.Leung. "The Bramble Cay melomys *Melomys rubicola* (Rodentia: Muridae): A First Mammalian Extinction Caused by Human-Induced Climate Change?" *Wildlife Research* 44, no. 1 (2017): 9–21.

Extinction due to climate change is more complex than one might think. For more on the process:

Cahill, A. E., M. E. Aiello-Lammens, M. C. Fisher-Reid, X. Hua, C. J. Karanewsky, H. Yeong Ryu, G. C. Sbeglia, F. Spagnolo, J. B. Waldron, O. Warsi, and J. J. Wiens. "How Does Climate Change Cause Extinction?" *Proceedings of the Royal Society B: Biological Sciences* 280, no. 1750 (2013).

Durance, I. and S. J. Ormerod. "Evidence for the Role of Climate in the Local Extinction of a Cool-Water Triclad." *Journal of the North American Benthological Society* 29, no. 4 (2010): 1367–1378.

Beever, E. A., C. Ray, P. W. Mote, and J. L. Wilkening. "Testing Alternative Models of Climate-Mediated Extirpations." *Ecological Applications* 20, no. 1 (2010): 164–178.

Wiens, J. J. "Climate-Related Local Extinctions Are Already Widespread among Plant and Animal Species." PLOS *Biology* 14, no. 12 (2016): e2001104.

PHOTOGRAPHY
AND ILLUSTRATION
CREDITS

Pages 1, 3, 5, 18, 64, 116, 186, 212: Clara Prieto

Pages 6, 240: Shutterstock/Anton Balazh, NASA

Page 11: NASA/Apollo 17 mission

Pages 14–15: Alexander von Humboldt, courtesy of the Biodiversity Heritage
Library collection

Page 17: Naia Morueta-Holme et al. "Strong Upslope Shifts in Chimborazo's Vegetation over Two Centuries since Humboldt." Proceedings of the
National Academy of Sciences of the United States of America, vol. 112,
no. 41, 2015, pp. 12741–12745.

Pages 22–23: NASA Earth Observatory images by Joshua Stevens, using
GEOS data from the Global Modeling and Assimilation Office at
NASA GSFC

Pages 24–25, 67, 120–121, 190–191, 228–229: Filippo Vanzo

Pages 30–31, 68: Matthew Maury, courtesy of the Library of Congress,
Geography and Map Division

Page 33: National Geographic Image Collection/staff

Pages 34–35: NASA, staff

Page 36: Point B Studios. "Wind Map" by Martin Wattenberg and
Fernanda Viégas

Pages 38–39, 54–55: Adventures in Mapping

Page 41: William C. Woodbridge, courtesy of Harvard University

Pages 42–43: Heinrich Berghaus, courtesy of Princeton University Library,
Treasures of Rare Book Division

Page 45: NASA Earth Observatory images by Joshua Stevens, using GEOS-5 data from the Global Modeling and Assimilation Office at NASA GSFC and MODIS data from NASA EOS-DIS/LANCE and GIBS/Worldview

Page 46: NASA Earth Observatory image by Joshua Stevens, using data from the Level 1 and Atmospheres Active Distribution System (LAADS) and Land Atmosphere Near real-time Capability for EOS (LANCE), and Landsat data from the US Geological Survey

Pages 48–49: NASA Earth Observatory image by Lauren Dauphin and Robert Simmon, using Landsat data from the US Geological Survey

Pages 50–51, 52, 60–61: Brian Buma

Pages 56–57: Daniel Huffman

Pages 68–69: artist not known, courtesy of the Library of Congress, Geography and Map Division

Page 71: James Poupard and Benjamin Franklin, courtesy of the Library of Congress, Geography and Map Division

Pages 72–73: National Geographic Image Collection/Ryan Morris and Alexander Stegmaier:

Pages 74–75: National Geographic Image Collection/Ryan Morris, Alexander Stegmaier, and John Tomanio

Page 77: National Geographic Image Collection/Ryan Morris

Page 79: David Liittschwager

Page 81: National Geographic Image Collection/Theodore Sickley

Pages 82–83: NASA Earth Observatory images by Joshua Stevens, using MODIS data from LANCE/EOSDIS Rapid Response, sea-surface temperature data from Coral Reef Watch, and

modified Copernicus Sentinel data (2018) processed by the European Space Agency

Pages 84–85: Samuel Augustus Mitchell, courtesy of the Library of Congress, Geography and Map Division

Page 86: Robert Szucs

Pages 89, 90: US Army Corps of Engineers

Pages 92–95, 152–153: Daniel Coe, Washington Geological Survey, Washington State Department of Natural Resources

Pages 98–99: NASA Earth Observatory images by Joshua Stevens, using Landsat data from the US Geological Survey

Page 101: G.K. Gilbert et al, courtesy of Great Basin Historical Society & Museum

Page 102: National Geographic Image Collection/Jerome Cookson

Page 105: NASA Earth Observatory images by Joshua Stevens, using IMERG data from the Global Precipitation Mission (GPM) at NASA/GSFC and National Water Information System data from the US Geological Survey

Page 107: David Rumsey Map Collection/California Irrigation Commission

Page 109: Charles H. Widdows, courtesy of the Holt-Atherton Special Collections, University of the Pacific Library

Page 110: courtesy of the Holt-Atherton Special Collections, University of the Pacific Library, and the California Department of Water Resources

Pages 112–113: Vincenzo Maria Coronelli; Wikimedia Commons, Released into the Public Domain

Page 114: Chris Brackley/Canadian Geographic, based on work by Dr. David Allan at the University of Michigan School for Environment and Sustainability

Pages 122–123: MDPI/ Daniel Clewley, Jane Whitcomb, Mahta Moghaddam, Kyle McDonald, Bruce Chapman, Peter Bunting. 2015. "Evaluation of ALOS PALSAR Data for High-Resolution Mapping of Vegetated Wetlands in Alaska." Remote Sens. 7, no. 6: 7272–7297.

Pages 124–125: National Geographic Image Collection/ Matthew Chwastyk

Pages 126–127: City of Phoenix Water Services Department

Pages 128–129, 155, 170–171: National Geographic Image Collection/Jack Unruh

Page 131: NASA Earth Observatory image by Joshua Stevens, using ICESat-2 data courtesy of Amy Neuenschwander (University of Texas) and Kaitlin Harbeck (NASA Goddard Space Flight Center), and Landsat data from the US Geological Survey

Pages 132–133: William H. Brewer, courtesy of the Library of Congress, Geography and Map Division

Pages 134–135: Bill Rankin, www.radicalcartography.net

Pages 138–139: National Geographic Image Collection/ Jerome Cookson, Greg Fiske, Theodore Sickley

Page 141: National Geographic Image Collection/ Samantha Welker

Pages 142–143: National Geographic Image Collection/ Clare Trainor

Page 145, 217: Scott Reinhard

Page 147: Benjamin Louis Eulalie de Bonneville and S. Stiles, courtesy of the Library of Congress, Geography and Map Division

Pages 148–149: John Lambert, Isaac I. Stevens, Selmar Siebert, courtesy of the Library of Congress, Geography and Map Division

Page 151: Courtesy of the Library of Congress, Geography and Map Division, and US Geological Survey

Page 157: NASA Earth Observatory/Expedition 57 astronaut crew

Pages 158–159: Josiah Gregg, courtesy of University of Texas Arlington Library Special Collections

Pages 162–163: Courtesy of the Library of Congress, Geography and Map Division

Page 165: Sean D. Willet, Scott McCoy, and Helen Beeson. 2018. "Transience of the North American High Plains Landscape and its Impact on Surface Water." Nature 561, 528–532.

Pages 166–167: NASA Earth Observatory map by Lauren Dauphin, using data from Jillian Deines et al. (2017).

Pages 172–173: National Geographic Image Collection/Sean McNaughton, Lisa Ritter, and Hiram Henriquez

Page 174, top: NASA/Expedition 58 astronaut crew

Page 174, bottom: Brian Buma

Page 177: Sir John Ross, courtesy of the Library of Congress, Geography and Map Division

Pages 178–179: Henry Overton, courtesy of the Library of Congress, Geography and Map Division

Page 181: National Geographic Image Collection/ NGM Maps

Page 183: NASA Earth Observatory map created by Jesse Allen, using data provided by David Olefeldt, University of Alberta

Page 189: William Hubbard, courtesy of the Library of Congress, Geography and Map Division

Pages 194–195: National Geographic Image Collection/Sean McNaughton, M. Brody Dittemore, and Lisa Ritter

Pages 198–199: National Geographic Image Collection/ John Tomanio

Pages 202–203: William Trent Rossell and James Loring Lusk, District of Columbia Engineering Department, courtesy of the Library of Congress, Geography and Map Division

Page 206: Descartes Labs

Page 208: NOAA, climate.gov, based on data provided by Portland State SUPR Lab

Pages 210–211: Treepedia, MIT Senseable City Lab

Page 215: Frans Lanting Photography, Lanting.com

Pages 218–219: H. E. Winzler and P. R. Owens, Purdue University; and Z. Libohova, USDA, NRCS, and National Soil Survey Center

Page 221: New Mexico Land Conservancy

Pages 222–223: Brad Stratton, The Nature Conservancy

Page 225: Historia de Gentibus Spetnetrionalibus by Olaus Magnus; released into the Public Domain

Page 227: National Geographic Image Collection/Staff

Page 234: Scott R. Loarie, Philip B. Duffy, Healy Hamilton, Gregory P. Asner, Christopher B. Field, and David D. Ackerly. 2009. "The Velocity of Climate Change." Nature 462, 1052–1055.

Page 239: Anna Eshelman

Page 243: Philippe Buache, courtesy of Osher Map Library, University of Southern Maine

Pages 244–245: Guillaume de L'Isle and Claude Berey, courtesy of the Library of Congress, Geography and Map Division

ACKNOWLEDGMENTS

Science is the process of standing on the shoulders of giants, looking a little bit farther, and reporting back. This work would not have been possible without the immense struggles and hardships of early explorers and scientists, often laboring in obscurity and difficult conditions. Although their names are arcane to most, individuals like Humboldt, Matthew Fontaine Maury, and Hayden transformed our view of the world. The frequency of satellite imagery can make it difficult to grasp how transformative the view they brought to the world actually was—a view of the earth as a whole, as a system-of-its-own-scale, beyond the human experience. Although this reality—that we are but small bits in a much larger environmental system—is still finding purchase with some, those intellectual and geographic explorers made the initial leap. This book is deeply indebted to them, but just as much to the archivists that have made those maps and datasets available via museums and the internet. People like Erin George, of the University of Minnesota Elmer L. Andersen Library, made the Glacier Bay section possible via her work facilitating access to old, analog datasets for interested researchers like myself. If not for her work, and the work of the libraries around the world (such as the US Library of Congress), much of our history would be lost. And while there are many things science can do, it cannot recreate the past for observation—once it is gone, it is gone.

Along those lines, we should not only thank the national leadership that made our views of the world possible, or the scientists that published the work, but also acknowledge the unknown Russian and US electrical engineers, mechanics, custodians, and generally uncredited team members that made satellites and satellite imagery possible, which allowed us to finally visualize the world as a whole. The US took the first picture of Earth from orbit on the Explorer 6 mission in 1959 (in 1946, rocket engineers had strapped a black and white camera to a V2 rocket and captured a grainy sub-orbital image), and the first true satellite images of the moon were likely taken by the Russian Luna 3, also in 1959. The astronauts of Apollo 17, captained by Eugene Cernan, first revealed Earth to the public via the "Blue Marble" in 1972 (mentioned in the introduction) and made the scale of our world's problems visceral. More important for environmental science, the Landsat mission was launched that year by visionaries such as Valerie Thomas; it provides the longest continuous satellite imagery record of Earth available. Today we enjoy a huge variety of imagery encompassing the entire natural world, but it started with Thomas, Cernan, and the workers on those awe-inspiring projects.

There is a huge debt, clearly, to the scientists that built the body of understanding and research underlying this book. Folks like Paul Hennon, who taught me so much about yellow cedar, and Vladimir Romanovsky, who arranged a literal journey into a real permafrost tunnel—it looks exactly as described in the text. Agencies like the National Oceanic and Atmospheric Administration, NEON, the European Space Agency, and others provided a wealth of information. Groups like National Geographic and the National Science Foundation provided the funding for much of the research highlighted here. Although I've included a suggested reading section, not all the science that is mentioned can be included—there is simply too much. My job was taking incredibly detailed work and distilling and placing it into the global-scale story of our world. Any mistakes made in the syntheses of their works are mine, and I sincerely apologize for any oversimplifications. I urge all readers to take the next step and visit a local university to ask questions, or search Google Scholar for their favorite topic, and then to contact those scientists directly. They deserve all the credit.

Today we have a wealth of data—almost too much to comprehend. Satellite imagery in a host of spectra, dense sensor networks, remote weather stations, careful experimentation in labs around the world, and smartphone-based community science observations stack onto the growing historical record. To make sense of that requires talent in data distillation, identifying the important and displaying it with clarity. A special thanks goes out to academic scientists with a flair for the visual, and who graciously allowed their expertise and/or imagery to be used in the communication of this story, from reviewing drafts to providing figures: Jason Amundson, Sean Willet, David Olefeldt, Anna Eshelman, and Daniel Clewley. Artists are now also data scientists, and talented creators such as Daniel Huffman, John Nelson, Scott Reinhard, Sarah Gilman, Filippo Vanzo, and Bill Rankin (among many others) need to be recognized—not just for their artistic gifts but for their skill and practice in choosing what data to show and what to leave out. Clarity and simplicity and accuracy is not easy, as I have found over and over with this writing.

Finally, bringing a visual book (or any book) to life requires a whole team of builders. Will McKay had the initial vision to start the project, providing the all-important bridge into the publishing process and frequent nudges and suggestions on early drafts. Adrianna Sutton lent her visual expertise and skills in finding talent to fill visual holes. Julie Talbot edited, translating my often (but inadvertent!) jargony language into more palatable phrasings and putting up with endless tweaking on my part. Clara Prieto, the artist who perhaps provided the most to the feel of the book through her cover and chapter openers, was crucial. A book of this kind—a mash-up between science text, data visualization, and cartography—needs an inspired visual theme to hold it together, which she provided.

This book brushes just the surface. The scale of the world is too large. But hopefully it is a gateway for the reader to continue exploring.

INDEX

BRIAN BUMA is a professor at University of Colorado, in Denver, in the Department of Integrative Biology, tracking ecological change in a variety of climates and ecosystems. He also holds an affiliate professor position at the University of Alaska, Fairbanks. His writing and research have been featured in the *New York Times*, *Los Angeles Times*, *Science* magazine, *Earth Island Journal*, and *High Country News*, among other outlets. He is also a National Geographic Explorer and a fellow of the Explorer's Club, and has led research expeditions that span the globe, embracing the adventurous side of science. Recent projects include an expedition to *Isla Hornos* (Cape Horn) in southern Tierra del Fuego in search of the world's southernmost tree, rediscovering and resurrecting a century-old backcountry Alaskan glacial plant study, exploring earthquake-triggered Himalayan landslides in Nepal, and establishing high-altitude research sites in Alaska and Colorado.